From Congo to GONGO

Rochelle Brock and Cynthia Dillard
Executive Editors

Vol. 116

Hope K. McCoy

From Congo to GONGO

Higher Education, Critical Geopolitics, and the New Red Scare

PETER LANG
New York · Berlin · Bruxelles · Chennai · Lausanne · Oxford

Library of Congress Cataloging-in-Publication Data

Names: McCoy, Hope K., author.
Title: From Congo to gongo : higher education, critical geopolitics, and the new red scare / Hope K. McCoy.
Description: New York, NY : Peter Lang, [2024] | Series: Black studies and critical thinking, 1947-5985 ; Volume 116 | Includes bibliographical references.
Identifiers: LCCN 2023038975 (print) | LCCN 2023038976 (ebook) | ISBN 9781636671734 (hardback) | ISBN 9781636671741 (paperback) | ISBN 9781433196027 (pdf) | ISBN 9781433196034 (epub)
Subjects: LCSH: Education, Higher–Social aspects–Africa. | Education, Higher–Social aspects–Russia (Federation) | Geopolitics–Africa. | Geopolitics–Russia (Federation) | Soft power (Political science)–Africa. | Soft power (Political science)–Russia (Federation) | Africa–Foreign relations–Russia (Federation) | Russia (Federation)–Foreign relations–Africa.
Classification: LCC LC191.98.A35 M33 2024 (print) | LCC LC191.98.A35 (ebook) | DDC 378.6–dc23/eng/20230920
LC record available at https://lccn.loc.gov/2023038975
LC ebook record available at https://lccn.loc.gov/2023038976
DOI 10.3726/b21182

Bibliographic information published by the Deutsche Nationalbibliothek.
The German National Library lists this publication in the German National Bibliography; detailed bibliographic data is available on the Internet at http://dnb.d-nb.de.

Cover design by Peter Lang Group AG

ISSN 1947-5985 (print)
ISBN 9781636671741 (paperback)
ISBN 9781636671734 (hardback)
ISBN 9781433196027 (ebook)
ISBN 9781433196034 (epub)
DOI 10.3726/b21182

© 2024 Peter Lang Group AG, Lausanne
Published by Peter Lang Publishing Inc., New York, USA
info@peterlang.com - www.peterlang.com

All rights reserved.
All parts of this publication are protected by copyright.
Any utilization outside the strict limits of the copyright law, without the permission of the publisher, is forbidden and liable to prosecution.
This applies in particular to reproductions, translations, microfilming, and storage and processing in electronic retrieval systems.

This publication has been peer reviewed.

Dedication
For my sister Faith

FULL BOOK ABSTRACT

In 1964, former Senator Fulbright stated: "Education is in reality one of the basic factors of international relations...quite as important as diplomacy and military power in its implications for war and peace" (Fulbright, 1964). Later, in the late 1980s, Joseph Nye famously coined the term "soft power", describing the ways that countries co-opt rather than coerce, often through culture and higher education (Nye, 1990). What is missing from these perspectives is the agency of those thought to be the target of influence. This book adds to and extends the literature on politics and power, using higher education as an example of the ways both nation states *and* individuals make choices and seek partners on the geopolitical stage. With a focus on Africa and Russia, this book traces the history of contact between the two regions, from the era of the Tsars, through the reign of the USSR, to the current agenda of the Russian Federation. During each time period—higher education, politics, and Black studies are woven together to create a patchwork of knowledge, each era with shifting values and purposes that influence diplomatic relations between Africa and Eurasia.

The qualitative methodology for this research included analysis of (N=394) archival documents, content analysis of (N=1,000) google reviews of GONGO-sponsored Russian Cultural Centers in Africa, in-depth interviews with African students and alumni of Soviet-era educational programs, and focus group and observation methods.

Utilizing a hybrid framework of neocolonialism, value homophily, transnational social capital, and critical geopolitics, this book offers vignettes centered on Egypt, Ethiopia, and Zambia, teasing out the various effects of Arab socialism, Marxist-Leninist socialism, and Indigenous Afrosocialism that impact both the culture and politics of these countries. (Keller, 1984; Pitcher & Askew, 2006; Torrey & Devlin, 1965).

With an emphasis on the legacy of Cold War programming through present day initiatives, this book not only analyzes the impact of higher education on geopolitical power, but also discusses the way cultural identities impacts migration patterns, school choice, and ultimately, global citizenship.

Keywords: Transnationalism, African Education, Russian Education, International Higher Education, Politics in Education, Soft Power, Diplomacy, Cultural Centers, Geopolitics

CONTENTS

	List of Illustrations and Tables	xiii
	Acknowledgements	xv
Chapter 1	African—Russian Relations: Historical and Political Contexts	1
	Introduction	2
	Notes on Reflexivity	3
	Education as a Diplomatic Tool	4
	Engaging Individuals	5
	Establishing Institutions	6
	Implementing Policies	7
	Contextualizing the Global Education Arena	8
	Theoretical Considerations	9
	Neocolonialism, Neoliberalism, and Soft Power	10
	Value Homophily	13
	Transnational Social Capital	13
	Critical Geopolitics	16
	Conclusions	17
	References	19

Chapter 2	Dissolving Borders: Transnationalism and the African Elite	25
	Introduction	26
	What is in a Name?	27
	Early African Russian Relations	27
	The Cold War and Soviet Partnerships	29
	Red Shadows: The Dark Side of Global Soviet Engagement	34
	Current Russian Higher Education Development in Africa	37
	Symbolic Partnerships	37
	New Hybrid Universities	39
	Scholarships to Russian Universities	42
	From Congo to GONGO: Russian Cultural Centers in Africa	44
	Closing Thoughts	46
	References	47
Chapter 3	Research Considerations and Theoretical Applications	51
	Introduction	52
	Higher Education Development in Africa during the Era of the Soviet Union	52
	Higher Education Development in Africa during the Era of the Russian Federation	54
	Development of the African Elite	55
	In Egypt	55
	In Zambia	56
	In Ethiopia	60
	Dignity, Political Camaraderie, and Respect	61
	Research Methodology	67
	Data Collection Procedures	68
	Archival Research	69
	Content Analysis	70
	In-Depth Sociocultural Data	72
	Data Analysis	72
	Limitations	74
	Positionality	74
	References	75
Chapter 4	The New Red Scare	79
	Introduction	80

The Purpose of Russian Cultural Centers in Africa 80
   State Perspectives 80
   Student Perspectives 84
Areas for Further Study 90
Conclusions and Implications 91
References 93

LIST OF ILLUSTRATIONS AND TABLES

Figure 1.	Theoretical Framework	11
Figure 2.	Data Sources	69
Table 1.	Google Reviews	71
Table 2.	Participant Demographics	73

ACKNOWLEDGEMENTS

The sincerest thank you to my mentors and colleagues: Drs. Walter Allen, Edmond Keller, Grant Parker, Andrey Rezaev, Parna Sengupta, Daniel Solórzano, Robert Teranishi, Rick Wagoner, and Fred Zimmerman. I truly appreciate your guidance and enthusiasm in my pursuit of interdisciplinary knowledge.

A heartfelt thank you to my family for all of your support: especially my sister Faith and my aunt Dolly.

A warm thank you to my close friends and colleagues: Anubha Anushree, Dionne Berry, Rahsaan Chionesu, Bartosz Chmielowski, Jason Cieply, Steve Coupet, Jennifer Daly, Betty Dao, Jennifer Hsu, Jen Furlong, Rich Gallion, Lally Gartel, Alexey Goloshchapov, Jay Harris, Ryan Hughes, Cameron Jones, Stacey June, Elizabeth Kalbers, Alex Khramova, Gina Mink, Jenny Lee, Adam Levin, Ameer Loggins, Nik Lund, Joellen McBride, Taylor Minas, Dayo Mitchell, Sara Mrsny, Rusana Novikova, Maro Oganesyan, Alissa and B. Pagels-Minor, Carlisle and John Rex-Waller, Nicole Robinson, Stefi Skuro, Azeb Tadesse, the Tertulia community, Averi Thomas-Moore, Yuri Zhayvoronok, Johana Zuazo, and мои русские друзья.

· 1 ·

AFRICAN—RUSSIAN RELATIONS: HISTORICAL AND POLITICAL CONTEXTS

Abstract: Prompted by Cold War-era political competition, the Soviet Union engaged in wide-ranging soft power initiatives on the African continent. This included supervising the building of key infrastructure, founding colleges, supporting the study and preservation of indigenous languages, and educating future political leaders in Africa. Although much of these programs were shuttered in the early 1990s, there has been a rekindling of interest in Africa since the mid 2000s, further catalyzed by the current political tensions between East and West, marked by the Russo-Ukrainian War.

This chapter introduces the topic of educational diplomacy and delves into Russia's current strategy: fostering symbolic partnerships, establishing hybrid universities, building cultural centers, and awarding scholarships to African students to study in Russian universities. Higher Education serves a diplomatic instrument: engaging individuals, establishing institutions, and implementing cross-national policies. Russian Cultural Centers represent a new tool in the diplomatic toolbox, and the result of a recent trend of prioritizing the non-university sector in African higher education.

Keywords: International Higher Education, Russian Education, African Education, Politics in Education, Soft Power, Diplomacy, Symbolic Partnerships, Cultural Centers, Geopolitics

Introduction

In 1958, 90-year-old W.E.B. Du Bois travelled to Russia to give lectures, attend events celebrating the anniversary of the October Revolution, and, to attend meetings with then Premier Nikita Khrushchev (Higbee, 1993). Du Bois had just recently received his passport, which had been confiscated by the United States government after years of political persecution. Like many African-American intellectuals of his time, such as Langston Hughes and Richard Wright–Du Bois was a long-term supporter of socialist ideals, and endured the combination of McCarthy-era paranoia, racism and oppression (Clark, 2016; Lewis, 1993; Morris, 2017).

During this trip (one of several to the Soviet Union in his lifetime), Du Bois met with Khrushchev to advocate for the development of an institute housed in the USSR that would focus on Pan-African academic research and cultural preservation (Morris, 1973; Weaver, 1985).

He arrived in Moscow on the heels of the 1957 World Festival of Youth and Students, which welcomed the largest number of African students that Russia had ever seen (Kotek, 2015). Two years later, Khrushchev convinced the Party to establish the Institute for African Studies, a research institution part of the Academy of Sciences, that remains active in Russia today (Rosen, 1963; Studies, 2021).

By 1960, the USSR founded Patricia Lumumba University (today known as RUDN University) in Moscow, an institution now famous for educating thousands of students from Africa, Asia, and Latin America (Rosen, 1973). A student festival and a single meeting between Du Bois and Khrushchev were the catalyst for a series of wide-ranging educational initiatives that targeted students from the newly liberated African nations of the 1950s and 1960s. With the decolonization of Sudan in 1956, Ghana in 1957, and Guinea in 1958—sprang up multiple programs for both technical training in African countries, as well as scholarship programs for African students to study abroad in the Soviet Union (Rosen, 1970).

By the time of the dissolution of the USSR in 1991, more than 400,000 African students had studied in Soviet-administered higher education institutions, including universities, military institutes, vocational schools, and short-term training programs. An additional 200,000 African students matriculated in Soviet-funded education programs based in their home countries (Patman, 2009). As of 1984, a total estimate of 45,075 of those students hailed from

Sub-Saharan African countries. From 1955 to 1984, the most scholarships were awarded to students from Ethiopia, followed by Nigeria, Madagascar, the Republic of Congo, Ghana, and Tanzania (Holt et al., 2014).

Considering the impact of these educational programs in Africa, the purpose of this book is to illustrate the nature of Russian higher education development on the African continent, including the scope, agenda, and relationships of power. By examining the historical context of these initiatives, one can consider how formal and informal postsecondary education shapes the geopolitical terrain, while predicting future diplomatic engagement. This book knits together the past and present political agendas playing out in international higher education.

Notes on Reflexivity

The research underpinning this book began with my first visit to Russia, in the spring of 2014, when I was invited to give a talk at the annual G8 Summit, held in Moscow. The timing was tumultuous, as this conference occurred mere weeks after Russia's annexation of Crimea. Due to Russia's actions, the Group of Eight (G8) ousted Russia from the forum, and the event was rebranded as a Global Summit. The energy during this trip was tense, but filled with excitement, with even my local tour guide proudly exclaiming how the annexation of Crimea was positive, a demonstration of Russia's power in protecting her people who lived in that region. Others were less enthusiastic, anticipating a future fraught with sanctions, war, trade embargos, and the return of Cold War-era hostilities of East vs. West.

Less than a year after my first trip to Russia, I returned and joined a research team in the Comparative Sociology department at St. Petersburg State University, traveling back and forth to consult and teach. Ultimately this led to a Fulbright grant in 2015, when I spent a year conducting academic research on how EU-initiated education policies impacted the ways that Russian faculty make meaning of their profession and the culture of education.

My experiences in Russia were not only academic. I also worked as a professional musician, singing jazz and playing piano in local cocktail bars and theaters, playing 60 concerts in one year, both solo and alongside local bands. These experiences in Russia—living, working, studying, creating music—inspired me to write my PhD dissertation on the impact of Russian education on the African diaspora. Myself, a Black American scholar moving across borders, existing transnationally between the pages of a contentious period in global history. The hope here in this book, is to expand the academic literature on African geopolitics.

I hope that the reader of this work will engage with ideas around the ways that power is wielded, and how students and political leaders in the African diaspora makes choices while moving through a world with increasingly permeable borders. Considering the current Russo-Ukrainian War, which began in 2014, reached a fever pitch in 2021, and continues through the publishing of this book in 2023—the topic of geopolitics, and role of the Russian state is a timely one. I invite educators, historians, foreign policy experts, and laypeople to reflect on the ways education and culture shape politics, using Russian higher education in Africa as a case study in soft power (Nye, 1990).

Education as a Diplomatic Tool

Although many of the Soviet-funded cross-national education programs benefitting African students were discontinued in the 1990s, the impact remains embedded in various aspects of society including public health, education, military, and politics. For example, as of 2014, over 11% of current practicing medical doctors in Ghana were trained in either the USSR or Russia (Holt et al., 2014). In addition, numerous African political leaders have degrees from Russian and Soviet institutions, including José Dos Santos of Angola, Ahmadou Touré of Mali, and Thabo Mbeki of South Africa.

The far reach of Soviet-era education programs in Africa, Latin America, and Asia, are an example of the inherent diplomatic influence of higher education on society. Not only a Soviet-era trend—China, France, Germany, and the United States all have extensive, successful education programming with similar goals. One example being the DAAD (German Academic Exchange Service), a state-funded scholarship program that not only funds postsecondary education of German students, but also provides funding to over two million international students in 1,800 programs, with the express purpose of promoting the German language and culture around the world. The DAAD also states that the organization "assists developing countries in establishing effective universities and advises decision makers on matters of cultural, education and development policy" (DAAD, 2021). This broad goal, specifically the latter portion on advising decision-makers—has deeper political implications, more so than merely sharing culture and language abroad.

These types of educational and cultural programs are examples of the concept of soft power (Nye, 2005). Soft power refers to the influence a country has over another, specifically excluding economic or coercive methods of persuasion. Developed by Morgenthau (1967), Knorr (1975), and Cline (1975), and most famously expanded on by Nye (1990)—soft power justifies the adoption and inclusion of policies and cultural norms in shaping the opinions of others. The language used to describe the impact of soft power is often aspirational and isomorphic, with countries adopting the norms of others, simultaneously seeking legitimacy (Altbach & Peterson, 2008; Cline, 1975; Knorr, 1975; Lo, 2011; Mattern, 2005; Nye, 1990).

Describing soft power, Nye delineates the difference from hard power, with a focus on co-opting rather than coercing others to gain the outcomes through three methods: culture, political values, and foreign policies. Nye argues that higher education fulfills all three of these methods, (1) through cultural exchange when international students come to the United States and live among American peers, (2) absorption of western political values in their coursework, and (3) positive foreign policies through opportunities to work together on research and other collaborative projects (Nye, 2005). Higher Education offers a wide range of opportunities to implement the tools of soft power. Some of these opportunities serve individuals—such as study abroad and research programs; some serve countries—such as international branch campuses and libraries; and other opportunities engage regions in systemic change via cross-national public policies (Altbach & Peterson, 2008).

Engaging Individuals

It is not only scholars and students that move individually across borders—but programs, education providers, and policies, shift as well. The higher education landscape is characterized by national governments increasingly seeking to drive internationalization, graduate employability, patterns in student mobility, as well as government and institutional initiatives to increase funding streams (Manners & Whitman, 2013). Approximately four million students study abroad each year, and a substantial population of faculty and scientific researchers participate in transnational collaborative opportunities across borders (Cantwell, 2019).

Former Senator Fulbright once stated "Education is in reality one of the basic factors of international relations . . . quite as important as diplomacy and military power in its implications for war and peace" (Fulbright, 1964). The 1946 establishment of the Fulbright Program specifically addressed a need for developing "post-war leadership" and "advancing mutual understanding" between countries (Fulbright, 1964). This series of exchange programs has often been described as tool of soft power—influencing and advancing the agenda of the United States throughout the world (Altbach & Peterson, 2008; Nye, 2005; Trilokekar, 2010).

Although much has been published on the individual impact of such study abroad programming like Fulbright, citing social and character development opportunities (Lee & Rice, 2007; Nye Jr, 2010; Scott, 2000)—the impact on institutions and state actors, is more nuanced. With a variety of program types targeting students, educators, and artists, Fulbright has provided funding to 370,000 people since its inception. This range of influence has included political leaders –37 heads of state, as well as 84 Pulitzer Prize winners and 59 Nobel Prize winners (Arndt & Rubin, 1993). Funneling transnational leaders into positions of power, Fulbright is and was a U.S. program with a focus on the Global, that significantly impacts the Local.

Establishing Institutions

The Fulbright program is only one of many soft power-oriented programs administered by the former United States Information Agency (USIA). In the 1960s, the USIA founded exchange programs such as Arts America, the robust broadcasting initiative: Voice of America, the Washington File information sharing service, and over 100 libraries around the world, through the United States Information Service (USIS) (Maack, 2001; "Role of libraries in the USIA program," 1961). The USIS libraries were a cultural branch of the USIA, established in 1953, charged with the responsibility of building an understanding of the culture, institutions of the United States with people around the world (Anderton, 1967). Although their expressed goal was to provide public access information for developing countries, many scholars argued that at best, these were institutions supporting U.S. propaganda goals, and at worst –the USIS libraries were adjuncts of the Central Intelligence Agency (CIA) in foreign countries, used as fronts to conceal surveillance and tracking activities (Anderton, 1967).

The expansion of international branch campuses is another soft power trend in transcultural higher education. There are currently over 200 of these institutions operating on a global scale, with the United States serving as the most significant provider, currently hosting 80 institutions around the world (Verbik, 2007). Branch campuses are entities owned wholly or partially by foreign providers that operate under the goodwill of the sending institutions, offering complete academic programs that result in degrees awarded by main the campus (Brauch, 2017).

Branch campuses are not a new concept; between 1962 and 1973, Massachusetts Institute of Technology (MIT) was instrumental in the establishment of the Institute of Technology (IIT) in India, the Aryamehr University of Technology in Iran, as well as the Indian Institute of Technology and Science (Scott, 2000). Michigan State University also has a foreign presence in India, as well as in Dubai, China, Burundi, Mozambique, Zambia, Tanzania, Kenya, and Mali, coordinating independent student research projects, and providing American students with an international context to their own education. However, these programs have expanded beyond serving foreign markets, with the benefits of these relationships sometimes serving the western side of the partnership, more so than that of eastern stakeholders (Trilokekar, 2010).

Implementing Policies

In addition to the individual and institutional venues of soft power, cross-national policies in education dictate even more extensive, long-term soft power influences. The Bologna Process (BP) is a prime example, a pan-European educational policy initiative that encourages the mobility of students and academic professionals. The BP eliminated the regulations that stunted this goal in the past, while streamlining accreditation policies and standardizing degree structures. A series of policy actions with the purpose of creating a European Higher Education Area (EHEA), the BP was born of the Bologna Declaration, an agreement initiated in 1999, signed by the Ministers of Education of 47 countries (Adelman, 2009). The purpose of the BP was to make European education more attractive, competing with the United States in recruitment of international scholars, research and innovation, and global university rankings. What began as an educational reform geared towards mobility, has expanded into multiple strategic initiatives, including working groups on lifelong learning, quality assurance methodologies, and employment development (Huisman & Van Der Wende, 2004).

The BP encouraged faculty and researchers to be mobile, engaging in cross-country partnerships. However, global university rankings privilege the elite universities, with scholars only choosing to partner with the most esteemed institutions, which are often located in Western Europe (Voegtle et al., 2011). While the objective of these types of cross-national policies encourages both faculty and students to be mobile—studying, working, and researching in institutions outside of their home country, this shift has been remarkably dominated by the movement of East to West (Heinze & Knill, 2008).

Contextualizing the Global Education Arena

Altbach and Peterson (2007) argue that the ability of higher education to transcend national boundaries, as well as the type of education programs these institutions are involved in, can significantly influence global relationships on a grand political scale (Altbach & Peterson, 2007). The rising trend in higher education has been to engage across borders and institutions, including dual degree programs, faculty consulting contracts, collaborative research projects, and joint degrees among other education programs (Roberts, 2009). Higher Education is changing to meet the needs of the future; new modes of education have progressed in response to globalization, engaging a variety of types of students and institutions.

The reach of digital initiatives such as Massive Open Online Courses, the expansion of branch campuses, and the massive movement of international students challenge traditional notions of brick and mortar higher education, as well as the need for, or value of the Humboldtian model of higher education, which necessitates liberal arts values and a four-year degree (Anderson, 2020). Along with new modes of learning are innovations in cross-national partnerships, continuing the tradition of education and diplomacy by expanding the borders of institutions, countries, and their subsequent ideologies and agendas. To better understand the present and future of education and diplomacy, this project studies a new institution type that by its very nature, crosses the borders of categorization, and the borders of nation states.

This is a new type of cross-national partnerships in higher education: Russian Cultural Centers (RCCs) in Africa. Much like their Chinese counterparts, RCCs share the goal of spreading their respective languages and cultures around the world. However, China's Confucius Institutes are also affiliated with local universities, whereas RCCs (in many, but not all cases) exist as separate, autonomous organizations.

RCCs also perform some of the same duties and tasks as traditional colleges and universities, including skills training such as computer technology and mathematics courses, social engagement such as networking and alumni events, and resources and support for African students to apply for Russian universities. I argue that these RCCs are an example of a fifty-year trend that prioritizes the educational development of the non-university sector in Africa, rather than efforts to support formal higher education (Omwami, 2013).

Theoretical Considerations

I frame my analysis of the purpose of Russian Cultural Centers through two perspectives: the State and the Student. The State includes Russia, Egypt, Ethiopia, and Zambia. Although these diverse state actors have different goals and historical contexts, they all wield political power over the Student. I define the Student as African alumni of educational programming developed by the USSR or Russia, current students enrolled in RCCs in Africa, and the general African public—which includes potential RCC students. When the State engages in neocolonial behaviors coupled with neoliberal pressures, the result is Soft Power, the concept famously theorized by Joseph Nye.

Moreno et al (2018) also link soft power and neocolonialism, using the example of the European Neighborhood Policy (ENP) enacted by the European Union, which unduly influences other countries both in terms of cooperation on human rights issues (soft power) and economic aid (neocolonialism) (Moreno et al., 2018). Von Eschen (2006) in her book about the U.S. State department-hosted jazz tours of the 1950s and 1960s, connected the cultural exchange mission of these tours to the pursuit of oil and uranium, stating that these concerts were both an attempt to bolster diplomatic relationships, but also were a cover for the American acquisition of natural resources in Africa and the Middle East (Von Eschen, 2000).

Antwi-Boateng (2017) also connects the concepts of soft power and neocolonialism, highlighting the uneven power relationship between China and Africa. Antwi-Boateng describes the pursuit of raw materials and unscrupulous trade agreements as neocolonial pressures, strategically combined with a soft power rhetoric of altruism, an increase in Chinese university scholarships for

African students, and the broadcast of the China Central Television network across 47 countries in Africa (Antwi-Boateng, 2017).

As discussed earlier, soft power is a method of getting what one wants without coercion. Countries seek to wield power and influence by making themselves appear attractive. Therefore, soft power is a strategy, and higher education is a tool of that strategy. Joseph Nye's concept of soft power includes three areas of influence: culture, political values, and foreign policies—all of which can be found in higher education development (Nye, 2005). Russian Cultural Centers are a type of informal, unaccredited higher education designed to lead to formal higher education.

I argue that soft power combined with capitalistic neoliberal values serves as a catalyst in the creation of new institutions like Russian Cultural Centers (RCCs). This idea supports Edith Omwami's identification of a fifty-year trend of state and supranational powers divesting from African universities in favor of private sector educational organizations (Omwami, 2013).

From the perspective of the Student, I theorize that students enrolled in RCCs may experience value homophily between their own culture and Russian culture. In addition, their experiences in RCCs and possible transfer to Russian universities develops students' transnational social capital, building a network across borders. These values and experiences cultivated in RCCs lead to the development and improvement of African Human Capital. Both the State and the Student engage in Critical Geopolitics, their behaviors and motivations infinitely impacting one another.

Neocolonialism, Neoliberalism, and Soft Power

The perspective of the State is informed by the theory of neocolonialism—first posited by former Ghanaian president Kwame Nkrumah in the 1960s. This theory refers to the cultural, economic, and political pressures used to control other countries, most often occurring between formerly colonized nations and their previous colonizers (Nkrumah, 1967). Dependency is at the heart of this lens, with countries from the Global North creating a system of exploitation that undermines the stability of newly liberated nations in the Global South (Baldwin, 1993). This exploitation occurs through tactics such as development aid and predatory lending which leads to cycles of unpayable debt (Gordon, 1997), or monopolizing natural resources at the expense of ecological damage, such as the European Union overfishing in West Africa, among other

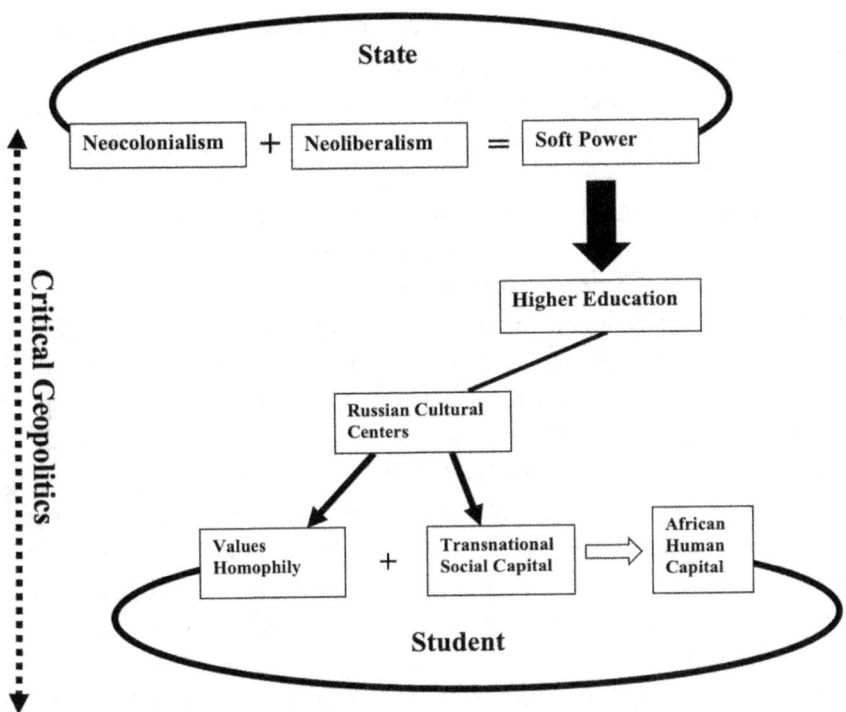

Figure 1. Theoretical Framework

tactics (Rodney, 2018). Nkrumah argued that the liberation and establishment of African nation states of the 1950s and 1960s was merely symbolic, due to the exploitative system of dependency that continued post colonialism (Nkrumah, 1967).

Although Russia never joined the desperate Scramble for Africa in the 1800s, and in fact was instrumental in decolonization efforts of the past century—I argue that the Federation still engages in neocolonialism as their goals have shifted along with their country's political leanings, from socialist to a capitalist oligarchy, much like the United States. Despite the USSR's vehement vocal opposition to European imperialism, the Russian Federation exhibits the third type of imperialism—a sphere of influence, in rekindling it's educational and cultural goals in the 2000s while simultaneously seeking access to the valuable natural resources on the African continent. Echa (2013) echoes this sentiment offering the examples of Ethiopia and Liberia—both countries were never colonized, yet due to the precarious nature of their economic system coupled with a reliance on foreign capital, they too suffer under the manipulation of neocolonialism (Echa, 2013).

What sets soft power apart from neocolonialism is the lack of economic influence, as Nye (1990) includes economics into coercive hard power, along with military might (Nye, 1990). However, Robert Young (2016) in his many writings on postcolonialism, has stated that "development" no longer only refers to economics, but now, due to globalization, also refers to culture, gender, politics, and societal norms (Young, 2016). In his seminal writings on neocolonialism, Nkrumah leans into a Marxist critique, highlighting the class conflicts inherent in postcolonial societies. He argues that as the elite ruling class of postcolonial countries focus their attention on gaining power and prestige from their former masters, the advancement of education and the eradication of poverty among the lower class is all but forgotten. It is this forgotten space of education which is fulfilled by the tools of soft power. In addition, as the educational training provided by RCCs to African students is not pro bono, and charges students individual tuition, economic motivations and power arrangements are present in addition to the sociocultural implications of transnational education exchange.

Altbach (1971) applies a neocolonial lens to the field of education, identifying the use of foreign textbooks, teachers, and practices in developing nations. The destruction of indigenous concepts of knowing, teaching, and learning, in favor of elitist normative practices led to a reliance on a western education to advance even within one's home country. In addition, the use of European languages in former colonies as the professional and governmental lingua franca is a stark example of neocolonialism in education (Altbach, 1971). Altbach also points to the reliance on foreign aid, foreign curriculum, foreign instructors, and foreign branch campuses (Altbach, 2014) which I argue can be identified in the proliferation of RCCs as feeder institutions into Russian universities.

When neocolonial behaviors are combined with neoliberal pressures, the result is the State engaging in soft power, a post-history strategy of geopolitics. In addition, those same neoliberal pressures have led to the fifty-year trend Omwami identified of investment in private educational enterprises in Africa, of which I argue RCCs are a prime example. Neoliberalism is a market-oriented approach to economics, pushing towards a free market approach to all aspects of society from education and social services to food distribution and trade. Much has been written about the globalized world, and neoliberal pressures in higher education. Rhoades and Slaughter (1997) use the phrase academic capitalism to refer to the past five decades of changes in university life, from student as learner, to student as consumer (Rhoades & Slaughter, 1997; Slaughter, 2004; Slaughter & Leslie, 1997).

Value Homophily

The perspective of the Student is guided by a framework of value homophily and transnational social capital. Homophily is a sociological theory most famously described as "love of the same" or "birds of a feather flock together" (McPherson et al., 2001). People bond with, and develop networks with others who share similarities, whether it be in level of education, political opinion, religious belief, or cultural background. Merton (1996) delineates between the different types of homophily, including: status homophily, which are the traits people are born with such as their race and ethnicity, and value homophily, which are the traits people develop, their beliefs and opinions, and worldviews (Merton, 1996).

The theoretical framework for this research is informed by value homophily, in describing the incentives students have to join RCCs, as well as the behaviors and practices that occur after attending RCCs. With value homophily, people feel justified in their beliefs due to the proximity of like-minded individuals (Huston & Levinger, 1978; Knoke, 1990). Marx and Spray (1972) observed the ways value homophily impacts how patients choose therapists, with religious background gleaned (or assumed) from doctors' professional biographies, and ethnicity. Combined, patients would make decide whether doctors might share their values and provide treatment acceptable to their worldview (Marx & Spray, 1972). However, Rauwolf et al found that in communities that share value homophily, there is a propensity for dishonesty, with the group less likely to point out fallacies amongst those they share beliefs. (Rauwolf et al., 2015).

Transnational Social Capital

The topic of transnational social capital was not originally included in the proposal of this research. However, the concept arose from the data analysis, cropping up in interviews with current RCC students and in the content analysis of RCC social media posts. One of the purposes of Russian Cultural Centers in Africa is to serve as feeder institutions to universities in Russia, offering the language classes necessary for academic study, but also serving as normative agents, legitimizing African applicants to Russian colleges. Although students may apply for scholarships even if they have not attended an RCC, the impressive placement rate, and access to the instructors in RCCs who provide insider knowledge on the application process are incentive to attend the institutions.

Social Capital is a sociological theory arising from the theories Aristotle, Durkheim, and most famously from Bourdieu, referring to the resources gained from networks of relationships among people (Bourdieu, 2011). From the intermittent interactions with our neighbors that form weak social ties to the deep bonds between families, to the professional contacts between colleagues—the strength of a network benefits the individuals in that network by providing more than just socioemotional connections (Granovetter, 1973). Social capital has economic implications, with job opportunities often found within a network, as well as romantic unions leading to marital mergers connected by a network. Inclusion in an influential network is a reflection of power, cooperating across ideologies and identities (Coleman, 1988; Granovetter, 2018; Sanders & Nee, 1996).

Transnational Social Capital refers to the networks that develop across borders, which is a necessity, and a competitive edge in an increasingly globalized world. Levy et al (2013) highlight the benefits of transnational social capital in business: having friends and colleagues in various regions gives a person advance notice of new trends and opportunities, whether it be a house not yet on the market, or a need in a local community that could benefit from a trade agreement (Levy et al., 2013). Companies can expand overseas when there is a local network available to ease their transition. Knowing which officials' palms to grease goes a long way in understanding the minutiae of foreign tax laws and other bureaucratic barriers.

In higher education, the topic of transnational social capital has been discussed at length, with students from similar cultural backgrounds serving as "bonding social capital" and students from different backgrounds serving as "bridging social capital", with these diverse bridging ties found to be empirically most dominant (Schartner, 2015). The literature also frequently examines how the social ties developed in college leading to postgraduate employment opportunities (De Graaf & Flap, 1988; Seibert et al., 2001).

Even more interesting—Moon and Shin (2019) found that when international students study in non-English speaking countries, they are often segregated into engaging with other international students, rather than their new domestic peers, leading to the development of transnational social capital in an unexpected way (Moon & Shin, 2019). Fincher and Shaw (2009) came to the same conclusion, identifying international student housing as a significant impact on social networks, segregating foreign students from both their peers and the local community (Fincher & Shaw, 2009).

These interactions, between students going through study abroad and being the Other, bonded in this experience with others. I argue this would be play out in the same way for African students studying in Russia. The Zambian and the Ethiopian student might come from different cultures, but being Black abroad supersedes these interethnic conflicts, while simultaneously creating new transnational networks in Africa, upon return to their respective home countries post-graduation. However, Waters and Leung (2013) in a study of foreign students enrolled in British university branch campuses, found that non-local degree programs have less of an impact cultivating group solidarity, describing a social disconnect (Waters & Leung, 2013). I argue this effect is less prescient in RCCs, due to the addition the value homophily and critical geopolitics unique to this particular phenomenon.

Language, or: knowledge of more than one language is a form of transnational social capital. Even without knowing people within a network, just the act of learning a new language provides the opportunity to develop a network, opening up more job opportunities, and chances for migration. However, even monolingual Americans enjoy the transnational nature of English serving as a lingua franca in business and diplomacy, the legacy of the imperialist British Empire. Language is centered in the mission of Russian Cultural Centers, even for the locations that don't serve as feeder institutions. Many RCCs located in former socialist republics serve as social spaces, for entertainment and connection between Russian speakers. The RCC in Ethiopia is also more nostalgia-oriented, providing a place for alumni of soviet-era programs to practice their forgotten Russian, and for Ethiopian scholars to seek translation services for their Aramaic-written publications.

In addition to the language aspect of transnational social capital, is the role networks play for international students in higher education. For students interested in studying abroad, having a friend or family member who studied in that country provides both a sense of comfort and safety. A transnational network also provides insider knowledge on what local customs to be aware of, what academic expectations are, and how to navigate both a foreign university and a foreign country (Alfred, 2013; Markley, 2011; Sanders & Nee, 1996; Waters, 2009). In the field of migration studies, this phenomena is referred to as cumulative causation theory, positing that as more members of a community decide to immigrate to a particular region, more will be encouraged to join them, now that family and friends have tested the waters and determined them to be safe and sustaining (Fussell, 2010; Liang, 2014; Maier, 1985; Massey, 1990).

The data from this research also suggests that African students who wish to study abroad are likely to return to their home country upon graduation, in order to apply the knowledge and skills that they developed in college, to their country's infrastructure and development. Thwarting concepts of brain drain and brain gain (human capital flight), these behaviors fall within the concept of brain circulation, referring to the circular movement of workers across borders (Blachford & Zhang, 2014; Robertson, 2006; Schmitt & Soubeyran, 2006). I argue that this brain circulation leads to a transnational identity that serves and supports the needs of the State, falling under the pressures of neo-colonial influence. Yet, I also argue this transnational identity simultaneously challenges the prior dominance of the West, engaging in Critical Geopolitics.

Critical Geopolitics

In the study of geopolitics, the focus is on how place impacts people and policies. Traditional modes of inquiry include examining how geography impacts political power, studying natural resources, spatial networks, and boundaries. Critical geopolitics, a poststructuralist approach, challenges assumptions made about the world based on previous notions of superpowers and spheres of influence. It is "a critique of dominant representations of international politics" (Sharp, 2013). Also called subaltern geopolitics, referring to the voices not usually heard in international relations and political geography discourse (Craggs, 2018; Sharp, 2011, 2013).

I offer two main critiques of the dominant discourse in the field of international education research. The first critique challenges the supposition that African students aspire to study abroad in the United States or the United Kingdom, decentering those regions due to a changing political landscape, accompanied by value homophily and transnational social capital. The second idea I turn over upon its head, is the idea of African countries passively acted upon by "greater" world powers. Instead, I offer examples of how both the African members of the State and the Student act with agency, both engaging in savvy political maneuvering, with the express goal of improving and expanding African human capital.

I move beyond a top-down approach to understanding the impact of cross-national policy. It is true that Russia, as a state is executing soft power with political agendas in creating these institutions, and African countries are engaging in soft power of their own, by *allowing* Russia (and other nations) to build these cultural centers in the first place. But another lens to consider is that African students who participate in these programs are doing so for their own reasons, they are not simply being "acted upon." African students are creating their own learning experience for various reasons: opportunities for economic mobility, and/or the chance to develop social and cultural capital through study abroad opportunities. Savage (2016) alludes to the niche I pursue in this work:

> The opening of the Soviet archives in the early 1990s has lent new perspectives on the dynamics of educational diplomacy, revealing that, rather than passively accepting these gestures of solidarity or soft power, the recipient States often negotiated and determined the terms of Soviet assistance and educational aid. (Savage, 2016, p. 35).

This is a unique approach to neocolonialism, as Russia (formerly socialist, but now an oligarchy), is engaging in the same behaviors they denigrated, not long ago. However, I do not subscribe to a Manichean model of thinking, not all is defined in terms of good and evil, oppressor and oppressed. In this instance, the hybridization of Russia's neocolonial and soft power activities serve as an alternative to the Western dominance of the United States and the United Kingdom.

Conclusions

Given fluctuating tensions with the West due to social conflicts and economic sanctions levied by Europe, the U.S., and Canada, the Russian government is rekindling past relationships. For example, since 2008, Russia has funneled funds, weapons, and military training to the Democratic Republic of Congo (Pham, 2010). In addition, each year since then, 50 military officers from the DR Congo have trained in military programs in Russia. In 2011, the Russian Minister of Culture traveled to Senegal to announce a substantial increase in scholarships for international students, in addition to establishing a Russian Cultural Center in Senegal (Kulkova & Sanusi, 2016). The RCC established in Kinshasa in the Democratic Republic of Congo, is physically located in the Diplomatic Academy, spatially connecting education and diplomacy in time. Pham (2010) highlights the geopolitical tensions here:

> Talk of a "new Cold War" may be premature, but it should not be forgotten that, during the original Cold War, Africa was a major theater of the Soviet Union's

competition, not only with the United States, but with the People's Republic of China. And while Beijing's burgeoning engagements across Africa have received considerable attention, the Kremlin's reemergence as a significant power in Africa has gone largely unnoticed, unwittingly giving an increasingly assertive Russia a free hand in forging multiple economic, political, and military ties. (Pham, 2010)

At the core of this phenomena is the ways that higher education is both subject to, and a tool of the pressures of neocolonialism. Lo (2011) argues that necolonialism is transnational by nature, distinct from colonialism with a flow of power from country-to-country, neocolonialism relies on the actions of transnational companies and actions (Lo, 2011). Lo uses Nye's concept of Soft Power as an alternative to neocolonialism, whereas I argue they are one and the same. Alatas (1993) warns of the dangers inherent in the exchange of ideas between the Global North and South, using the term "mental captivity", to refer to a reliance on Western thought and ideas, leading to academic dependence at the expense of indigenous knowledge (Alatas, 1993).

Another topic to consider is that in the field of international and comparative education research, there is often a focus on Europe, Asia, and the United States. This homogeneity of sources, ideas, and practices can lead to stagnation. Methodological nationalism, according to Wimmer and Schiller (2003, p. 576) is the process and focus on the primacy of the Nation-State as the focus of research and analysis in the social sciences; and is a common reductive practice in comparative higher education (Wimmer & Schiller, 2003). Building this bridge between Eurasia and Africa, may rekindle former networks, identifying potential areas of cross-national research collaboration.

This work also addresses issues of power, to better understand future partnerships, and to expose western audiences to the machinations of higher education outside our own borders. Education is a vehicle of mobility with individual and societal motives, both to the benefit and the stability of the Nation State. However, in an increasingly interconnected world, the scope of education extends beyond the national. Higher education is a multinational entity that shapes economic, political, and social development over time. This book fills a gap in the literature on African diplomacy from an interdisciplinary standpoint by blending education, international relations, sociology, and political science literature.

In an increasingly interconnected world, the lens of the global expands the scope of education beyond the national. Examining this topic helps us to understand the changing landscape of international education, expands perspectives, and serves as a case on globalization, and the impact of capitalism on peripheral Soviet actors.

When reading the following chapters, reflect on the following four key areas:

- **The Student:** African student motivations to study abroad in Russia (vs. "the West"), as well as the global networks African students develop and the power they wield after studying abroad.

- **The State:** How and why Russia partners with African governments and engages in critical geopolitics, and how higher education is a diplomatic tool at the center of this examination.

- **Policy:** How specific African-Russian education initiatives are linked to economic and military agreements, and how those agreements lead to the control of, and access to natural resources.

- **Culture:** How shared values bring seemingly disparate cultural groups together, towards a shared vision of the future.

References

Adelman, C. 2009. The Bologna Process for U.S. Eyes: Re-learning Higher Education in the Age of Convergence. Washington, DC: Institute for Higher Education Policy. Location reference: www.ihep.org/Research/GlobalPerformance.cfm.

Alatas, S. F. (1993). A Khaldunian Perspective on the Dynamics of Asiatic Societies. SSRN Electronic Journal. https://doi.org/10.2139/ssrn.2650607

Alfred, M. V. (2010). Transnational migration, social capital and lifelong learning in the USA. International Journal of Lifelong Education, 29(2), 219–235.

Altbach, P. (1971). Education and Neocolonialism. Teachers College Record, 72(4), 543–558.

Altbach, P. G., & Peterson, P. M. (2007). Higher Education in the New Century: Global Challenges and Innovative Ideas (Vol. 10). Sense Pub.

Altbach, P. G., & Peterson, P. M. (2008). Higher Education as a Projection of America's Soft Power. Soft Power Superpowers: Cultural and National Assets of Japan and the United States, 37–53.

Anderton, L. D. (1967). USIS Libraries: A Branch of USIA. Peabody Journal of Education, 45(2), 114–120.

Arndt, R. T., & Rubin, D. L. (1993). The Fulbright Difference: 1948–1992. Transaction Publishers.

Antwi-Boateng, O. (2017). New World Order Neo-Colonialism: A Contextual Comparison of Contemporary China and European Colonization in Africa. Journal of Pan African Studies, 10(2), 177–195.

Baldwin, D. A. (1993). *Neorealism and Neoliberalism: The Contemporary Debate*. Columbia University Press.

Blachford, D. R., & Zhang, B. (2014). Rethinking International Migration of Human Capital and Brain Circulation. *Journal of Studies in International Education, 18*(3), 202–222. https://doi.org/10.1177/1028315312474315

Bourdieu, P. (2011). The Forms of Capital. (1986). *Cultural Theory: An Anthology, 1*, 81–93.

Brauch, N. (2017). Bridging the Gap. Comparing History Curricula in History Teacher Education in Western Countries. In *Palgrave Handbook of Research in Historical Culture and Education* (pp. 593–611). Palgrave Macmillan UK.

Cantwell, B. (2019). Are International Students Cash Cows? Examining the Relationship between New International Undergraduate Enrollments and Institutional Revenue at Public Colleges and Universities in the US. *Journal of International Students, 512*, 512–525.

Clark, K. (2016). The Representation of the African American as Colonial Oppressed in Texts of the Soviet Interwar Years. *The Russian Review, 75*(3), 368–385.

Cline, R. S. (1975). *World Power Assessment: A Calculus of Strategic Drift*. Boulder, CO: Westview Press.

Coleman, J. S. (1988). Social Capital in the Creation of Human Capital. *American Journal of Sociology, 94*, S95–S120.

Craggs, R. (2018). Subaltern Geopolitics and the Post-Colonial Commonwealth, 1965–1990. *Political Geography, 65*, 46–56. https://doi.org/https://doi.org/10.1016/j.polgeo.2018.04.003

DAAD. (2021). *Who We Are*. German Academic Exchange Service. Retrieved July 18 from https://www.daad.org/en/about-us/who-we-are/

De Graaf, N. D., & Flap, H. D. (1988). "With a little help from my friends": Social Resources as an Explanation of Occupational Status and Income in West Germany, The Netherlands, and the United States. *Social forces, 67*(2), 452–472.

Echa, N. (2013). The Historical Conjuncture of Neo-Colonialism and Underdevelopment in Nigeria. *Journal of African Studies and Development, 5*(5), 70–79.

Fincher, R., & Shaw, K. (2009). The Unintended Segregation of Transnational Students in Central Melbourne. *Environment and Planning A, 41*(8), 1884– 1902.

Fulbright, J. W. (1964). *Old Myths and New Realities and Other Commentaries*.

Fussell, E. (2010). The Cumulative Causation of International Migration in Latin America. *The ANNALS of the American Academy of Political and Social Science, 630*(1), 162–177. https://doi.org/10.1177/0002716210368108

Gordon, L. R. (1997). *Her Majesty's Other Children: Sketches of Racism from a Neocolonial Age*. Rowman & Littlefield.

Granovetter, M. S. (1973). The Strength of Weak Ties. *American Journal of Sociology, 78*(6), 1360–1380.

Heinze, T., & Knill, C. (2008). Analysing the Differential Impact of the Bologna Process: Theoretical Considerations on National Conditions for International Policy Convergence. *Higher Education, 56*(4), 493–510.

Higbee, M. D. (1993). A Letter from WEB DuBois to his Daughter Yolande, Dated "Moscow, December 10, 1958" Introduction and Footnotes. *The Journal of Negro History, 78*(3), 188–195.

Holt, J., Newhouse, S., & Ukhova, D. (2014). *Scholarships and the Healthcare Human Resources Crisis: A Case Study of Soviet and Russian Scholarships for Medical Students from Ghana.*

Huisman, J., & Van Der Wende, M. (2004). The EU & Bologna: Are Supra-and International Initiatives Threatening Domestic Agendas? *European Journal of Education, 39*(3), 349–357.

Huston, T. L., & Levinger, G. (1978). Interpersonal Attraction and Relationships. *Annual Review of Psychology, 29*(1), 115–156.

Knoke, D. (1990). *Political Networks: The Structural Perspective.* New York, NY: Cambridge University Press.

Knorr, K. E. (1975). *Power of Nations.* Basic Books.

Kotek, J. (2015). *Students and the Cold War.* Springer.

Kulkova, O. S., & Sanusi, H. A. (2016). Russia-Ghana Relations in the Past and the Present: A Time-Proven Partnership. *Vestnik RUDN. International Relations, 16*(2), 296–310.

Liang, Y. (2014). The Causal Mechanism of Migration Behaviors of African Immigrants in Guangzhou: From the Perspective of Cumulative Causation Theory. *The Journal of Chinese Sociology, 1*(1). https://doi.org/10.1186/s40711-014-0002-6

Lee, J. J., & Rice, C. (2007). Welcome to America? International Student Perceptions of Discrimination. *Higher Education, 53*(3), 381–409. https://doi.org/10.1007/s10734-005-4508-3

Levy, O., Peiperl, M., & Bouquet, C. (2013). Transnational Social Capital. *International Journal of Cross Cultural Management, 13*(3), 319–338. https://doi.org/10.1177/1470595813485940

Lewis, D. L. (1993). *WEB Du Bois, 1868–1919: Biography of a Race.* Macmillan.

Lo, W. Y. W. (2011). Soft Power, University Rankings and Knowledge Production: Distinctions between Hegemony and Self-Determination in Higher Education. *Comparative Education, 47*(2), 209–222. https://doi.org/10.1080/03050068.2011.554092

Maier, G. (1985). Cumulative Causation and Selectivity in Labour Market Oriented Migration Caused by Imperfect Information. *Regional Studies, 19*(3), 231–241. https://doi.org/10.1080/09595238500185251

Manners, I., & Whitman, R. (2013). Normative Power and the Future of EU Public Diplomacy. In *European Public Diplomacy* (pp. 183–203). Springer.

Marx, J. H., & Spray, S. L. (1972). Psychotherapeutic "Birds of a Feather": Social- Class Status and Religio-Cultural Value Homophily in the Mental Health Field. *Journal of Health and Social Behavior, 13*(4), 413. https://doi.org/10.2307/2136834

Massey, D. S. (1990). Social Structure, Household Strategies, and the Cumulative Causation of Migration. *Population Index, 56*(1), 3. https://doi.org/10.2307/3644186

Mattern, J. B. (2005). Why Soft Power Isn't So Soft: Representational Force and the Sociolinguistic Construction of Attraction in World Politics. *Millennium, 33*(3), 583–612.

McPherson, M., Smith-Lovin, L., & Cook, J. M. (2001). Birds of a Feather: Homophily in Social Networks. *Annual Review of Sociology, 27*(1), 415–444.

Merton, R. K. (1996). *On Social Structure and Science.* University of Chicago Press.

Morgenthau, H. J. (1967). *Politics Among Nations: The Struggle For Power and Peace.*

Maack, M. N. (2001). Books and Libraries as Instruments of Cultural Diplomacy in Francophone Africa during the Cold War. *Libraries & Culture,* 58–86.

Markley, E.-M. (2011). Social Remittances and Social Capital: Values and Practices of Transnational Social Space. *Calitatea Vieţii, 22*(4), 365–378.

Moon, R. J., & Shin, G.-W. (2019). International Student Networks as Transnational Social Capital: Illustrations from Japan. *Comparative Education*, 55(4), 557–574. https://doi.org/10.1080/03050068.2019.1601919

Moreno, N., Puigrefagut, A., & Yárnoz, I. (2018). The European Union's Soft Power: Image Branding or Neo-Colonialism. *Center for Global Affairs & Strategic Studies. Working Paper* [WP-05/2018].

Morris, M. D. (1973). The Soviet Africa Institute and the Development of African Studies. *The Journal of Modern African Studies*, 11(2), 247–265.

Morris, A. (2017). *The Scholar Denied: WEB Du Bois and the Birth of Modern Sociology*. University of California Press.

Nkrumah, K. (1965). Neo-Colonialism, the Last Stage of Imperialism. London: Thomas Nelson & Sons, Ltd.

Nye, J. S. (1990). Soft Power. *Foreign Policy*, 80, 153–171.

Nye, J. (2005). Soft Power and Higher Education. *Forum for the Future of Higher Education* (Archives).

Nye Jr, J. S. (2010). The futures of American power-dominance and decline in perspective. *Foreign Aff.*, 89, 2.

Omwami, E. (2013). Non-University Sector Reform. *Journal of Comparative & International Higher Education*, 5(Spring), 5–8.

Patman, R. G. (2009). *The Soviet Union in the Horn of Africa: The Diplomacy of Intervention and Disengagement* (Vol. 71). Cambridge University Press.

Pham, J. P. (2010). Back to Africa: Russia's New African Engagement. In *Africa and the New World Era* (pp. 71–83). Springer.

Rauwolf, P., Mitchell, D., & Bryson, J. J. (2015). Value Homophily Benefits Cooperation But Motivates Employing Incorrect Social Information. *Journal of Theoretical Biology*, 367, 246–261. https://doi.org/https://doi.org/10.1016/j.jtbi.2014.11.023

Rhoades, G., & Slaughter, S. (1997). Academic Capitalism, Managed Professionals, and Supply-Side Higher Education. *Social Text* (51), 9. https://doi.org/10.2307/466645

Roberts, A. (2009). *Civil Resistance and Power Politics: The Experience of Non- Violent Action from Gandhi to the Present*. Oxford University Press.

Robertson, S. L. (2006). Brain Drain, Brain Gain and Brain Circulation. *Globalisation, Societies and Education*, 4(1), 1–5. https://doi.org/10.1080/14767720600554908

Rodney, W. (2018). *How Europe Underdeveloped Africa*. Verso Trade.

Role of Libraries in the USIA Program. (1961). *ALA Bulletin*, 55(2), 180–181.

Rosen, S. (1970). The USSR and International Education: A Brief Overview. *The Phi Delta Kappan*, 51(5), 247–250.

Rosen, S. M. (1963). *Soviet Training Programs for Africa* (Vol. 9) [Bulletin]. US Dept. of Health, Education, and Welfare, Office of Education.

Rosen, S. M. (1973). *The Development of Peoples' Friendship University in Moscow*. Office of Education, Institute of International Studies.

Sanders, J. M., & Nee, V. (1996). Immigrant Self-Employment: The Family as Social Capital and the Value of Human Capital. *American Sociological Review*, 231–249.

Savage, P. (2016). *Reading between the Lines: African Students in the USSR.*

Schartner, A. (2015). 'You cannot talk with all of the strangers in a pub': A Longitudinal Case Study of International Postgraduate Students' Social Ties at a British University. *Higher Education*, 69(2), 225–241.

Schmitt, N., & Soubeyran, A. (2006). A Simple Model of Brain Circulation. *Journal of International Economics*, 69(2), 296–309. https://doi.org/10.1016/j.jinteco.2005.06.011

Scott, P. (2000). Globalisation and Higher Education: Challenges for the 21st Century. *Journal of Studies in International Education*, 4(1), 3–10.

Seibert, S. E., Kraimer, M. L., & Liden, R. C. (2001). A Social Capital Theory of Career Success. *Academy of Management Journal*, 44(2), 219–237.

Sharp, J. (2011). Subaltern Geopolitics: Introduction. *Geoforum*, 42(3), 271–273. https://doi.org/10.1016/j.geoforum.2011.04.006

Sharp, J. P. (2013). Geopolitics at the Margins? Reconsidering Genealogies of Critical Geopolitics. *Political Geography*, 37, 20–29. https://doi.org/https://doi.org/10.1016/j.polgeo.2013.04.006

Slaughter S. & Rhoades G. (2004). *Academic capitalism and the new economy : markets state and higher education*. Johns Hopkins University Press.

Slaughter S. & Leslie L. L. (1997). *Academic capitalism : politics policies and the entrepreneurial university*. Johns Hopkins University Press. Retrieved October 22 2023 from http://edrev.asu.edu/reviews/rev14.htm.

Studies, I. f. A. (2021). *Russian Academy of Sciences*. Retrieved July 17 from https://www.inafran.ru/en/

Trilokekar, R. D. (2010). International Education as Soft Power? The Contributions and Challenges of Canadian Foreign Policy to the Internationalization of Higher Education. *Higher Education*, 59(2), 131–147.

Verbik, L. (2007). The International Branch Campus: Models and Trends. *International Higher Education*, 46., 14–15.

Voegtle, E. M., Knill, C., & Dobbins, M. (2011). To What Extent Does Transnational Communication Drive Cross-National Policy Convergence? The Impact of the Bologna-Process on Domestic Higher Education Policies. *Higher Education*, 61(1), 77–94.

Von Eschen, P. M. (2006). *Satchmo blows up the world: Jazz ambassadors play the Cold War*. Harvard University Press.

Waters, J. L. (2009). Transnational Geographies of Academic Distinction: The Role of Social Capital in the Recognition and Evaluation of 'overseas' Credentials. *Globalisation, Societies and Education*, 7(2), 113–129.

Waters, J., & Leung, M. (2013). A Colourful University Life? Transnational Higher Education and the Spatial Dimensions of Institutional Social Capital in Hong Kong. *Population, Space and Place*, 19(2), 155–167. https://doi.org/10.1002/psp.1748

Weaver, H. (1985). *Soviet Training and Research Programs for Africa*.

Wimmer, A., & Schiller, N. G. (2003). Methodological Nationalism, the Social Sciences, and the Study of Migration: An Essay in Historical Epistemology 1. *International Migration Review*, 37(3), 576–610.

Young, R. J. (2016). *Postcolonialism: An Historical Introduction*. John Wiley & Sons.

· 2 ·

DISSOLVING BORDERS: TRANSNATIONALISM AND THE AFRICAN ELITE

Abstract: Eurasia and Africa have long enjoyed diplomatic ties, both before and after colonial occupation in Africa, and throughout the various eras of political regimes in Eurasia. These relationships spanned military and economic development, political camaraderie, and educational initiatives, oftentimes when Western powers had no interest in the continent beyond predatory debt collection.

In the late 1500s when Russia fought off Turkish Muslim invasion and occupation, the ancient kingdom of Abyssinia (now Ethiopia) offered their aid to Ivan the Terrible, catalyzing centuries of diplomatic ties between Africa and Eurasia. This chapter weaves the current political agenda with the context of the past, tracing the various ideological agendas that catalyzed educational diplomacy between Africa and Eurasia.

This chapter analyzes the ways that Soviet and Russian-educated Africans wield geopolitical power. Offering vignettes of politicians from Egypt and Zambia, two major themes are discussed in this chapter: (1) the development of the African Elite, and (2) dignity, political camaraderie, and respect. Theories of neocolonialism and transnational social capital are applied to the topic of Eurasian-African relations, highlighting an often-cyclical shifting of power between the State and the Student.

Keywords: Russian History, Soviet History, Geopolitics, African Education, Neocolonialism, Social Capital, Higher Education, History, Diplomacy

Introduction

In the Luapula Province in northern Zambia there is a yearly ceremony celebrated by the Ushi tribe called the Chabuka. Chabuka, which means "crossing" in the Ushi language, celebrates their ancestors' migration from the Luba-Lunda Kingdom of Congo (what is now the Democratic Republic of Congo), into their current home of Zambia. The ancestors of the Ushi crossed the Luapula River at low tide, with the aid of the stones beneath the water guiding their feet to safety (Arts, 2020; Shumba, 2016; Tula, 2018).

Each year there are celebrations and cultural performances during the Chabuka ceremony, including a reenactment of the crossing, with people carrying heavy stones as they cross the water, while spectators reflect on the journey of their predecessors (Tula, 2018).

In this way, the ancestors were aided by the foundation of stones that guided them to a new land, and in turn their descendants carry the weight of stone in recognition of those who came before. Holding this image in mind, consider the rocky foundation of political agendas that inspired Zambian students in their crossing into Russia in the 1960s. Although most students returned home after completing their degrees in Soviet universities, the transnational power of their social capital and their elite status in society, deem borders more permeable than they might be for the working class.

Long before the arrival of Europeans, Africa enjoyed longstanding trading relationships with Asia and the Middle East. With economic trade came cultural exchange, from Ghanaian gold traders converting to Islam from interaction with their Berber-speaking business contacts—to Swahili Kingdom traders bringing Indian cuisine to the continent, *sambusas* only a few letters distant from *samosas*. In addition to economic prosperity and cultural exchange, education has an ancient grounding in the African continent that influenced European colonists, not the other way around, as western-dominant history suggests. From the tragically famous Library of Alexandria in 3rd century BCE Egypt, to religious study in Christian monasteries in 4th century Ethiopia, to Islamic mosque universities such as University of Al Quaraouiyine, founded in 859 CE, all pre-date the most prestigious of European institutions (Zeleza, 2006).

In this way, I ask the reader to challenge notions of afropessismism that relegate this topic to deficit thinking. This lack of knowledge of pre-colonial African culture often frames the experience of the Black Diaspora as forever flawed, and ultimately doomed (Zeleza, 2006). Although this work examines the impact of Russia educational programs in Africa, I remind my audience that higher education existed on the continent long before colonial or postcolonial foreign intervention.

In the 21st century, Russian Cultural Centers have sprung up as informal postsecondary institutions, charging students for an education while lacking accreditation and normative status. RCCs serve multiple purposes—political, cultural, and educational. To fully understand the context of the problem, this research inquiry offers an interdisciplinary examination of the literature, including examining political science publications, historical archives, arts and media archives, and education research. This chapter provides the historical context of the educational partnerships which paved the way for the establishment of Russian Cultural Centers on the African continent.

What is in a Name?

Throughout this work there are instances where the words: Soviet, Socialist, and Russian, are used often, but should not be thought of as interchangeable words. It is important to explain the distinction between these terms, and how they impact the analysis of the historical context and the research data. The names *Soviet Union* and *USSR* both refer to nation states. However, the word *Soviet* relates to power, specifically the political power wielded between the years of 1917 and 1991 by the confederation of republics who were members of the USSR/Soviet Union. *Socialist* relates to the ideological and political values guiding the policies implemented, and values internalized by the citizens of the USSR, for example: free education for all people. The term *Russian* refers to the Slavic people who make up 81 % of the population, but it also refers to the nation state of the Russian Federation. In a sentence: there are many Russians who are not *Russian*, but, are not foreigners either, with 19 % of the population consisting of 190 ethnic groups, who are not ethnically Russian, but are indigenous Eurasian peoples.

Early African Russian Relations

In the late 1500s when Russia fought off the invasion of Turkish Muslim occupation, the ancient kingdom of Abyssinia offered their aid to Ivan the Terrible. Later, in the 1600s, explorer Hiob Ludolf published a book on his travels, including a history of Ethiopia, in the Russian language. These early interactions led to geological surveys and gold mining by Russian explorers in the Wollega province (Yakobson, 1963). Russia's history in Africa further developed in the 19th century during the Tsarist regime.

The most esteemed African in Tsarist Russia was Abram Petrovich Gannibal, the son of a local ruler in Cameroon. Gannibal was abducted by Turks as a young boy and sold into slavery, later adopted and freed by Russian monarch Peter the Great in 1705. A skilled mathematician and military engineer educated in France, Sweden, and Russia, Gannibal built fortresses (while exiled in Siberia) in addition to his position as a military strategist and nobleman (Schmemann, 2010). Gannibal's eleven children were members of the Russian nobility, eventually producing the writer and poet Alexander Pushkin. Pushkin is to Russians, what Shakespeare is to the British, though many would argue Pushkin is even more revered than Shakespeare, as Russians have a famous saying: "Пушкин—наше всё" translating to: "Pushkin is our everything."

In his writing, Pushkin spoke often of his heritage and pride in his African phenotypic features. Gannibal's influence also led to Russia establishing relationships in South Africa, with trade routes in the Transvaal province in 1898, as well relationships in North Africa, building a consulate in Morocco, that same year (Coles, 1999).

The longest and most prominent relationship on the continent is with Ethiopia, with many Ethiopians also claiming ancestry with Alexander Pushkin, alleging his grandfather was from Ethiopia and not Cameroon (or Ghana or Central Africa, as other scholars speculate) (Coles, 1999; Demassie, 2021). Ethiopia is a country with which Russia shares historic cultural ties and religious camaraderie—both with a majority Orthodox population. Indeed, prior to the second Italo-Ethiopian war of the 1930s, Ethiopia was a deeply religious orthodox country, devoted to a social system similar to feudalism, much like Russia prior to 1917.

V.F. Mashkov's diplomatic meetings with Emperor Menelik I in 1889 established formal trade between the countries, bolstered by mutual respect of both sides' orthodox beliefs (Clarke, 2011). That mutual respect supplied the mountain guns that helped Ethiopia win the Battle of Adwa against Italian colonialism in 1895, leading to the official establishment of diplomatic relations in 1897 (Drake, 1987). Between 1901–1913 a number of Ethiopian soldiers and officers attended cadet school in Russia, including Tekle Hawariat, the future author of Ethiopia's constitution (Van Creveld, 1990). At the age of 12, Hawariat was adopted by an elite Russian family in 1897, with whom he shared a close relationship. Later colloquially named "the Raven Baron" and "Petya the Abyssinian," Hawariat lived in Russia for seventeen years. In many ways, Hawariat served as a child diplomat, satiating the Russian romanticization of his Ethiopian heritage while informally brokering relations between the two regions (*The Raven Baron*, 2021).

Despite the cultural respect and positive engagement, the Kremlin's marked interest in Ethiopia was also a strategy to exert influence on Egypt, impact the Nile region, and gain control of the Red Sea, while showing strength to their British rivals. Ethiopia was hypothesized to be a "peaceful penetration" of the interior of the African continent (Pham, 2010; Yakobson, 1963). This peaceful penetration was but a taste of the impending complex geopolitical web, the threads of which were spun a few short years later with the global conflicts that began in 1917.

The Cold War and Soviet Partnerships

Despite the necessity of an alliance with the USSR that allowed Allied forces to win World War II, the tensions of the Cold War began in 1917, when President Woodrow Wilson changed allegiances during Russia's civil war (Foglesong, 2014). Although the USSR and Western powers enjoyed positive relations in the 1930s-1940s, those relationships soured by 1947. After benefiting from the Soviet military might in World War II, altering a dangerous fascist course of history—the U.S. published the Truman Doctrine of 1947, catalyzing a forty-year Cold War, in an attempt to curb their geopolitical influence.

The Bolshevik Revolution and subsequent communist state threatened American ideals and imperial sovereignty. From 1917 through 1991, the Soviet Union and the United States engaged in a series of hostile engagements: the rapid expansion of atomic nuclear weapons, the Space Race, proxy wars, propaganda campaigns, and soft power-designed cultural and educational programs. Both regimes strategized and competed for hearts, minds, and geopolitical power (Gaddis, 2006). Understanding the purpose of Russian Cultural Centers in Africa today, requires grounding in the political context that shaped the landscape where these institutions have been allowed to flourish.

Although the United States and the USSR never engaged in an outright war, the damage from manipulative proxy wars caused millions of dollars' worth of destruction and incalculable human carnage and suffering in Afghanistan, Algeria, Angola, Bolivia, Cambodia, China, Cuba, the Czech Republic, Chad, the Democratic Republic of Congo, the Dominican Republic,

Eritrea, Ethiopia, Grenada, Guatemala, Indonesia, Iran, Italy, Kenya, Korea, Laos, Lebanon, Libya, Malaysia, Morocco, Mozambique, Oman, Paraguay, Saudi Arabia, Somalia, South Africa, Slovakia, Taiwan, Thailand, Vietnam, Yemen, and Zimbabwe (Gerőcs, 2019).

The anti-imperialist messaging and influence of Marxist-Leninist ideology catalyzed the rebel movements in Angola and Mozambique, and heavily influenced the political regimes in Ethiopia, Somalia, and Tanzania, at various stages in their early days of liberation (Russell & Pichon, 2019). Backed by the Soviet Union, Cuba also intervened in African political conflicts during the Cold War, most famously in the 1963 Sand War in Algeria and the Angolan Civil War in the 1970s, but also in Benin, Guinea, Egypt, Ghana, the Republic of Congo, and Mali.

The impact and aid was extensive, so much so that during a 1991 trip to Havana, President Nelson Mandela stated "we come here with the sense of the great debt that is owed the people of Cuba. What other country can point to a record of greater selflessness than Cuba has displayed in its relations to Africa?" Indeed, thousands of construction workers, doctors, nurses, soldiers and teachers travelled from Cuba to provide resources and support to African liberation movements throughout the Cold War, coupled with military and logistic support from Soviet partners (Schmidt, 2013).

In addition to meddling in the civil wars and violent border disputes of the Global South, the United States and the USSR strategized other methods of power, weaving espionage and surveillance into the goals of influencing citizens of developing nations undergoing political transition from the 1950s through the 1970s. The CIA was tasked with sabotaging the political infrastructure of opposing groups, covertly and overtly throughout developing countries (Saunders, 2013). Between 1957 and 1985, the USSR signed agreements with 37 African countries, while the U.S. was engaged in sabotaging these new regimes (Mcclellan, 1993).

However, beyond these coercive measures lies the influence of ideology, with economic support fueling sociocultural change. The political and ideological goals of the Soviet Union encouraged widespread partnerships in education. The USSR historically championed the internationalization of higher education, stemming from early socialist ideals. The Bolsheviks—leaders of the revolutionary working class, believed that the October Revolution of 1917 would lead to a Global Revolution, and subsequent New World Order. Education was the vehicle for the proletariat to advance their understanding of their current condition (Lilge, 1968). Advancing higher education was no longer about the Russian elite, but about educating all people, training workers, and disseminating the ideology necessary for the eventual transition from war communism to utopian communism (Kuraev, 2014).

Edu-political relationships were nurtured with newly liberated African nations, Central Asian countries, and Latin America. Coined the "Affirmative Action Empire" the international egalitarianism of Soviet education has been noted by historians (Martin, 2001). In education, the largest number of Soviet scholarships were awarded to Ethiopia, followed by Nigeria, Congo (Brazzaville), Ghana, Egypt, Madagascar and Tanzania (Mortimer, 1972).

The first African students began studying in Soviet universities in the 1920s, enrolling in Stalin Communist University of the Toilers of the East (KUTV) in Moscow. A new university, KUTV was originally tasked with training workers for the eastern borderlands, specifically training Arabs, East Asians, South Asians, Turks, and Jewish people. In 1923, the mission was adapted to include both African and African American students. The courses these early students enrolled include Russian language, Political Economy, History, Leninism, Military Science, and English language. And in addition to these seemingly innocuous studies, was a hidden curriculum on espionage, small arms training, coding, guerilla warfare, rule of conduct under surveillance, and interrogation techniques (Nash et al., 2016).

After the subsequent establishment of the Soviet Union in 1922, contact with Africa expanded beyond scholarship programs and enrollment in Soviet universities to fulfill specific needs of the State. In the 1930s, geological exploration became the priority in Soviet-African relations. The establishment of a research institute funded Soviet specialists who travelled to Angola, Benin, Ethiopia, and Mali, building laboratories and national geological centers. This later led to the construction of power plants and hydroelectric stations in the 1960s, such as the Malka Wakana in Ethiopia. The Aswan Dam built across the Nile River in Egypt was a Soviet project that took ten years to complete and cost millions of dollars. These engineering constructions required vocational training of local workers, which were taught by Russian instructors. In addition, Soviet specialists built schools and hospitals, as well as Bahir Dar University, an institution still respected today in Ethiopia (Kochetkova, 2009).

A boom of internationalization was triggered by the death of Stalin in 1953. Violently xenophobic, Stalin, in stark contrast to his predecessor Lenin, stood in the way of transnational opportunities. Khrushchev had the express goal of training personnel to increase the industrialization of former colonies, seeking political friendships in exchange for technical education to increase the numbers of pilots, medical experts, and construction workers in these countries (Nash et al., 2016).

On the heels of the 1959 establishment of the Institute for African Studies in Moscow, enrollment of African students in Soviet universities increased by 120 % between the 1959–1960 and 1960–1961 academic years, whereas the number of African students studying in the United States had only increased by 40 % those particular years (Weaver, 1985). Deterred by civil unrest and racist violence against African Americans in the United States at the time, African students were motivated by the opportunity for an education in the USSR. By 1960, there were 777 African students studying abroad in universities in the USSR, the majority of whom hailed from Egypt (at the time, called the U.A.R.), Guinea, Ghana, Ethiopia, Somalia, and Sudan, though there were also students from seventeen other African countries studying in the Soviet Union that year (Rosen, 1963).

The universities that hosted the majority of African students were Peoples' Friendship University (originally named Patrice Lumumba University, after the Congolese politician and martyr), Saint Petersburg State University (at the time named Leningrad State University), Shevchenko University (at the time Kiev State University), the National University of Uzbekistan (then known as the Central Asian University in Tashkent—which also founded its own preparatory academy specifically for students from Africa, Asia, and Latin America), and the Tashkent State Agrarian University (Rosen, 1963).

In addition to enrolling African students interested in teacher training, health services, science, and sports—universities also established indigenous African language programs for Russians interested in working abroad, as well as for African students working on the preservation of cultural and artistic artifacts. Leningrad State University offered language training programs in Amharic, Hausa, Luba, Luganda, Kikongo, Swahili, and Yoruba. These programs also required companion courses in "the cultures and economies of African peoples" (p.3) with the express goal that at the end of this series of courses, undergraduates would have working, conversational proficiency in two or more related African languages (Rosen, 1963).

At Moscow State University, students had the option of majoring in African languages and culture. The degree required six years (5,672 hours) of course work, language laboratory, practical training, and seminars. The program also required students to specialize in one western language (English, German, or French), and two eastern languages, one being Arabic, and the other usually Amharic or Hausa, the languages local to Ethiopia and northern

Nigeria respectively. These requirements were in addition to the expectation that students would be fluent in Russian, reading and conducting their examinations (both written and oral) in Russian. All students, both Russian and African, were required to take three courses of political nature: political economy, dialectical and historical materialism, and history of the communist party of the Soviet Union (Rosen, 1963).

In contrast to Western scholarship of the time, Soviet art history courses respected African national art tradition and valued the cultural specificity of the many forms of African art. Soviet art historians critiqued Western perspectives on their lack of respect and lack of nuanced historical knowledge of the richness of African artistic traditions. When African artists like Eshetu Tiruneh and Tedasse Mesfin moved to Moscow to study at the prestigious Surikov and Repin art academies, they were encouraged not to replicate Soviet art and historical figures, but to pursue projects of their own histories, traditions, and contemporary movements. Rather than collecting the art of Africans to display in museums, relegated to limited categories, the USSR valued and respected these cultural contributions, rather than demanding assimilation, as Western traditions required (Nash et al., 2016).

One of the first cultural and educational exchange programs in the 1960s were a series of short-term group tourism trips, funded by the USSR, lasting from to several weeks to several months. These trips would welcome young African students to tour the Central Asian republics, at the time famed for their industrial innovation, as well as the famous cities of Moscow and St. Petersburg. In 1960, 200 African students attended these short-term programs. By 1963, the USSR had initiated three-year technical training programs on the African continent in agriculture, mining, power plant maintenance, and wood and glass processing across Egypt (at the time, called the U.A.R.), Guinea, Ghana, Ethiopia, and Sudan (Weaver, 1985).

Decolonization offered the chance to connect and influence nations seeking independence. The Soviet government not only funded liberation movements, but also provided military training. In the 1960s alone, the Soviet Union had either financial or military involvement in the Angola Civil War, the Nigerian Civil War (of 1967–1970), and both sides of the Ethiopian-Somali war of the late 1970s (Tareke, 2009). Several prominent political leaders in Guinea, Ghana, Ethiopia, Angola and Mozambique were supportive of Marxist ideology and welcomed both the economic support, but also the opportunity for social change beyond the political regimes of their colonial oppressors (Mortimer, 1972). Considering the fact that the western supranational economic powers like the World Bank and the IMF directly and publicly opposed the development of higher education in newly liberated countries like Burkina Faso, and the United States offered limited exchange programs for young Africans, the USSR was primed to step in and fill the educational void (Nash et al., 2016).

Red Shadows: The Dark Side of Global Soviet Engagement

By the end of Khrushchev's reign in 1964, the USSR spent 25 % of its new initiatives budget on economic aid and programming for Africa. New premier Brezhnez, disappointed in failed alliances with Egypt, Ghana, Guinea and Mali reduced that expenditure to 10 %, after halting funding to regimes he considered "too unreliable to warrant major investments in their loyalty" (Nash et al., 2016). However, in 1974 funding was reupped as new nationalist movements in Africa sprung up that also supported the ideological goals of the USSR.

The Carnation Revolution in Lisbon initiated a transfer of power in Lusophone Africa to socialist groups in Mozambique, Guinea, and Cape Verde, and a coup in Addis Abba deposed the anointed Emperor Haile Selassie, whose Pan African philosophy and choice to avoid sides had blocked Soviet goals in the past. These two events suggested a new Marxist-Leninist alignment of ideals, prompting Brezhnez to tout the necessity and success of Soviet-led transnational education programs.

In 1981, the USSR was training 72,090 international students, including 34,805 from the Africa, in a variety of programming; by 1986, the Soviet Union had spent a total of 1.7 million U.S. dollars on foreign student scholarships. These education programs were not merely socialist training modules, in fact, they filled the gap left by colonial powers, such as the training of electricians to replace the Portuguese technicians who fled Mozambique in 1975 (Kotek, 2015). And yet despite the egalitarian ideals and positive contributions to African education, there were conflicts and outright hostilities that contradicted these partnerships.

> The lofty anti-colonial rhetoric of the Soviet establishment could not conceal the country's homegrown racism and its officially inspired xenophobia. Africans in the Soviet Union were often confused by the mind-boggling mixture of state-sponsored propaganda and the reality of everyday racism and the selfless generosity and the warmth they encountered in many Soviet people. (Matusevich, 2008)

The 1963 suspicious death of Ghanaian student Edmund Assare-Addo in Moscow was believed to be murder, due to his romantic relationship with a Russian woman student. After police failed to arrest anyone, international students and their Russian allies led a protest of hundreds of people that ended in a violent clash with Moscow police. It had been 40 years since a political protest was held at Red Square and was catalyzed by the civil rights movement efforts in the United States, with protesters holding signs that read: "Moscow, a Second Alabama." (Matusevich, 2008)

In addition to outright violence and hostility, African students experienced a syrupy display of forced kindness, with Russian colleagues and instructors fearful of Kremlin reprisal if they expressed racist attitudes which were strictly prohibited in an utopian socialist society (Matusevich, 2008). Cowcher (2016) offers an example in the 1975 children's book "First Time in Moscow" that was published and distributed throughout Ethiopia at the time. The book, written by Russian authors, describes a young African boy visiting Russia for the first time, marveling over the achievements of the culture, describing the exceptionalism of the State, respecting the legacy of Lenin, and expressing his gratitude to the USSR.

However, the book failed to impress Ethiopians, who, within their own culture, considered themselves to be exceptional in comparison to other African ethnic groups. This ham-fisted approach was insulting, as the child depicted in the book never references which country he is travelling from, to visit Russia, erasing his identity relegating him to a generic, and forgettable character (Nash et al., 2016). This patronizing attitude was reported by many African students in Russia, one Ugandan student stating in the 1960s "I was beginning to feel uncomfortable from all this flattery, which had a touch of condescension in it, too. I began to feel that this was racial discrimination, but as it were, in reverse" (Matusevich, 2008).

In addition to the patronizing attitudes and condescending assumptions visited on African students in the USSR, there was also a marked flip flopping of political allegiances, that these education partnerships were meant to nourish. For example, the USSR secretly supported Italy during the 2nd Italo-Ethiopian War of 1935, despite their outward support for the abolishment of colonial institutions (Patman, 2009). The back and forth shifting of Soviet support between Ethiopia and Somalia during the Ogaden war of the late 1970s demoralized many African allies, and racist conflicts on college campuses in Moscow led to a decreasing population of African students in the Soviet Union (Nash et al., 2016).

> During the period of reforms, generally known as *perestroika* and *glasnost*, ushered in by the last Soviet leader Mikhail Gorbachev, the Soviet press commentary on Africa grew increasingly negative. Both political commentators and people in the street often attributed the economic decline of the once-powerful Soviet Union to "too much aid for Africa". (Martone, 2008)

The Soviet focus on the needs of the State and the military-industrial complex garnered educational advancements not only in diplomatic strategy, but also in science, technology, engineering, and mathematics, which as result, developed a worldwide legacy of Russian research innovation (Smolentseva, 2003). However, the democratization and decentralization of public institutions led to a marked decline in higher education funding, resulting in the cancellation of many of the collaborative relationships with students in Africa (Rezaev, 2006).

In addition, the Russian people balked at Gorbachev's reforms, whose capitalist notions caused the system to collapse. The Soviet agenda of sending money overseas, and supporting foreign students, was no longer tolerated during a time of economic desperation.

By the 1980s, in contrast to Ethiopian artists who had previously been respected and valued at Moscow art academies, new student artists like Bekele Mekonnen were now mocked by their instructors when they presented their work. The students who arrived from various African countries, were often the children of elite families. Throughout the Cold War there was a stewing resentment among Russians living in impoverished conditions, observing wealthy bourgeois Africans enjoying special privileges in the USSR. The Soviet scholarship package included not only tuition and higher quality room and board, but also stipends for meals, travel reimbursements, and access to high level Party officials.

One student recounted the experience of sitting down and putting his expensive jacket onto his chair, to be approached by a young Russian woman who indignantly told him "I paid for your jacket" implying that the Soviet scholarships that brought him to her university to study, were excessive, funding fashion rather than supporting the global class struggle and class solidarity espoused by the State (Nash et al., 2016).

Issam Khouraj describes the desperation of living in Moscow during perestroika, stating:

> The changes on the street were so dramatic. The generosity, the kindness...suddenly the tension on the street was so evident...Being a foreign suddenly becomes an

obstacle rather than a celebration. It was one of the most desperate times I felt in my life, that is you don't know if you're going to have the next meal. Crime became evident which you'd never heard of before. Racism became evident. (Nash et al., 2016)

A Sri Lankan student echoes this sentiment, observing that Soviet racism extended beyond Anti-Blackness to include Asian students as well, stating (p. 32):

Many foreign students were targeted, I myself was beaten when I was in Kiev. We were accused of robbing their money; we were accused of taking their jobs; we were the scapegoats, by the time I left to Soviet Union. (Nash et al., 2016)

The combination of nationalistic policy changes, economic constraints, and long-simmering racism formerly hidden by socialist policies led to the end of 70 years of formal transnational education policies between Africa and Eurasia (Nash et al., 2016).

Current Russian Higher Education Development in Africa

In 2014, sanctions from the West in response to the annexation of Crimea led President Putin to ramp up the rekindling of old Soviet partnerships in Africa, leading to an increase in official visits, signed agreements, arms sales, and soft power activities, including in higher education. I identify four methods of higher education development currently pursued by Russia in Africa with varying purposes: (1) symbolic partnerships, (2) new hybrid universities in Africa, (3) scholarships to Russian universities, and (4) Russian Cultural Centers.

Symbolic Partnerships

I define three practices as symbolic partnerships: (1) the media-publicized diplomatic visits between regions, such as Russian Minister of Foreign Affairs' multiple trips to five countries in Africa in 2018; (2) conferences and conventions such as the massive Russia-Africa Summit co-hosted with Egypt in Sochi in 2019; and (3) written agreements and memorandums of understanding, that have been signed over the years that sometimes lead to actionable plans, but do not always bear tangible fruits.

In this term "symbolic partnerships" I combine the ideas of symbolic power (Bourdieu) and symbolic interactionism, from Mead & Cooley (Farberman, 1985). Symbolic power is institutional, requiring the complicity of those being dominated, recognizing their place in a social hierarchy. It refers to power wielded due to "renown, prestige, honor, glory, [or] authority." A common example is power and influence exerted within churches and schools (Bourdieu, 1979).

Symbolic interactionism focuses on the relationships between people, and the way that language and communication are used to understand the world. People choose to interact with what they find meaningful, and they ascribe that meaning to things, based on their interactions with others (Blumer, 1986). Combining these two concepts is an apt description of the partnerships influencing higher education in Africa.

These agreements begin as symbolic gestures, are followed up with media-publicized visits from Russian politicians, are later codified into action by private and State-run Russian corporations, which then leads to higher education development. Here, I offer an example from Ethiopia. Russia and Ethiopia signed a Memorandum of Understanding (MoU) in 2017, as a gesture of symbolic partnership. In 2018, the following year, Minister of Foreign Affairs Sergey Lavrov took multiple trips to the continent that year, visiting Angola, Zimbabwe, Mozambique, Namibia, and Ethiopia. In a 2018 interview published in an Ethiopian newspaper, Lavrov stated:

> We intend to have a detailed discussion…on ways of enhancing the bilateral cooperation with the emphasis on its trade, economic and investment component, implementation of joint projects, particularly in the energy sector, including nuclear energy. Among the promising areas is Russia's assistance to Ethiopia in building its own scientific and research capacities in developing basic and applied sciences. Specifically we plan to create an Ethiopian center for nuclear science and technologies based on a Russia-designed research reactor. (Contributor, 2018)

In this same interview, Lavrov gushes over the positive diplomatic relations between Ethiopia and Russia, harkening back to their shared cultural heritage in Orthodox Christianity. After teasing out this value homophily, Lavrov zeroed in on the goal of building a nuclear science center in Ethiopia, priming the public for the next stage of development.

Later, in October 2019, Ethiopia and Russia signed an intergovernmental nuclear arrangement, with the goal of developing Ethiopia's nuclear facilities within ten years. The arrangement was formalized during the Russia-Africa Economic Forum at Sochi, when Russian government-owned Rosatom Nuclear Energy Corporation signed an accord with the Ministry of Innovation and Technology of Ethiopia, to provide technical and technological supplies for atomic energy projects. It is important to note that the Russian government-owned Rosatom State Nuclear Energy Corporation is the largest nuclear company in the world, conducting business with South Africa, China, Egypt, Finland, India, and Turkey, amongst others (Fikade, 2019).

The MoU stated that Ethiopia would be responsible for the security, storage and safety regulations of nuclear materials, as well as establishing a central authority responsible for regulating "waste treatment, education and training with nuclear and radioactive materials and substances" (Fikade, 2019). In addition, Russia engaged in relationship building between Ethiopia and Rwanda on this new nuclear energy project, with both countries joining the International Atomic Energy Agency (IAEA). In the same news article about this MoU, Russia also announced it planned to forgive twenty billion dollars of Ethiopian debt. The connection between this MoU on nuclear energy and the development of higher education is clear, when considering this 2019 press release posted on the Rossotrudnichestvo website:

> The Ethiopian side reacted with great attention and interest to the proposals of the Russian side. It was agreed that the Ethiopian Ministry of Higher Education will consider the possibility of sending Ethiopian scholarship holders to study in Russian universities in the specialty in nuclear energy as early as the 2020/2021 academic year. (*The delegation of Russian universities in Ethiopia*, 2019)

Through the symbolic partnerships, marketing in the media, and diplomatic visits—this nuclear energy project is an example of how a simple Memorandum of Understanding leads to higher education development.

New Hybrid Universities

The Russian Federation's educational transnationalism also includes operating 41 branch campuses outside the country, 38 of which are located in former republics of the USSR (Chankseliani, 2021; Mäkinen, 2016). In addition, in the past decade Russian universities have begun offering instruction in the English language, even at the most prestigious of universities such as Saint Petersburg State University. Combining these two trends, Russia has now begun building new hybrid universities.

These new hybrid institutions seek to expand the transnational reach of a Russian education, while diversifying a new generation of scientists and scholars. One such institution in Africa is the Egyptian-Russian University (ERU) located in Badr City, with an additional campus in Suez. ERU was founded in 2006, as one of the stipulations of an official agreement of cooperation in education between Egypt and Russia, finalized during President Putin's visit to the country in 2005. Notably, it was President Hosni Mubarak, himself an alum of a Soviet education program, who signed the decree that established the university. Here, we see the legacy of the impact of Soviet education and the benefits of developing an elite class of Africans.

Providing some context on the higher education landscape in Egypt: there are currently 26 public and 31 private universities, with thirteen of these schools landing in the top 100 of the Times Higher Education global rankings (as of 2018). There are also hundreds of private technical schools, vocational programs, military academies, and for-profit institutions operating in the country, with 20% of Egyptian students studying at private institutions. Egypt serves as a major hub within the Middle East, with a very large population of international students from other MENA (Middle East North African) countries. ERU is not the only foreign institution, as there is an American University in Cairo, the German University in Egypt, the British University of Egypt, and the French University of Egypt. However, it is only ERU and the even more recently established Chinese Egyptian University, that act as hybrid institutions in partnership with Egypt, rather than outsiders operating branch campuses on foreign land (Cochran, 2012; Hartmann, 2008; Mohamed & Trines, 2019; Richards, 1992).

ERU focuses on STEM disciplines, housing four departments: the Faculties of Pharmacy, Dentistry, Engineering, and the Faculty of Management, Professional Technology and Computers. ERU is located on a 32-acre campus complete with dormitories, a food court, a library, and sports facilities. The language of instruction is English, a compromise between the local language of Arabic, and the language of the faculty instructors: Russian. A coeducational institution, the university offers bachelor's degrees, and markets itself as a place where students can prepare to attend graduate school abroad, either in Russia, or in other countries (*Egyptian Russian University*, 2021).

The establishment of ERU was not merely a development of higher education—it was the building blocks for Egypt's New Administrative Capital city (NAC), a 58-billion-dollar project that will serve as an international hub of Afrofuturistic innovation. Located 28 miles east of Cairo in the desert, the plan to build the NAC was announced in 2015, almost a decade after ERU was established in this exact geographic region (Loewert & Steiner, 2019). Egypt has also cultivated multiple international partners in the development of the NAC.

Immediately after the announcement, Egypt partnered with China State Construction Engineering Corporation (CSCEC) to build government offices, luxury hotels, high-rise apartments, and a convention center in the NAC (Ayembe, 2021). Note the connected battling higher education interests here, as the new Egyptian Chinese University (ECU) was founded two years prior to this construction agreement, in 2013. The completed NAC will be roughly the size of the country of Singapore and is also located halfway between the critically advantageous seaport of Suez and the historic and popular tourist destination of Cairo (Ayembe, 2021).

A major feature of this new city will be dedicated areas for educational institutions, a technology and innovation park, hospitals, and two massive religious feats: a cathedral and a mosque. Taking a page from the desert success of Dubai, the NAC will be a "smart city" with high tech amenities, as well as artificial lakes, skyscrapers, a massive theme park, solar energy farms and international airport. Germany has also stepped into the arena, signing an agreement in January 2021 to design an electric railway to Cairo, one of several new light rail projects being built in the NAC (Ayembe, 2021). A Belgian construction firm is working on designing a new Egyptian Museum, and a UAE based firm won a contract to handle all waste management and city cleaning in the NAC (Mathews, 2021).

Although Russia has expressed interest in working with Egypt on cyber security issues and information technology development in the NAC (*Egypt, Russia to cooperate in ICT sector, Africa,* 2019), they have mainly focused on adjacent projects, such as building Egypt's first nuclear power plant in the city of El Dabaa. Part of a larger plan called Egypt Vision 2030, the completion of NAC is expected by the year 2022, with numerous residential neighborhoods, skyscrapers, and business offices already constructed, as of July 2021. The new Ministry of Defense buildings have completed construction in the NAC. Called the "Octagon" the sprawling, futuristic architecture are a science fiction movie fan's wildest dreams (Abdelmoaty & Soliman, 2020; Ayembe, 2021; Bolleter & Cameron, 2021; Loewert & Steiner, 2019; Mathews, 2021).

It seems ERU will no longer be the main higher education institution in NAC. In November 2017, Egypt's Minister of Education announced there will be six new hybrid or branch campuses opening in the city, from Canada, France, Hungary, Sweden, the United Kingdom, and the United States. As the number of international students in Egypt has doubled in the past decade, Egypt has become one of the top three destinations for students in MENA countries, after Saudi Arabia and the United Arab Emirates (Mohamed & Trines, 2019). The expansion of the NAC has solidified Egypt as a coveted future destination of higher education, with Russia partially responsible for these advancements, by being the first to test the waters of hybridity by building Egyptian Russian University back in 2006.

In an interview with Alana, a twenty-five-year-old Egyptian-born woman based in Los Angeles, she offers a practical (or cynical) view of Russian higher education development in her home country:

> Some of my family had to move back [to Egypt] and I thought I might have to go, but my visa problems ended up finishing, so I stayed here. But if I went, I was looking into a master's degree, and maybe a scholarship to Russia. A lot of the programs teach in English because nobody wants to learn Russian. I don't want to live there, but they have excellent IT programs, I could get the diploma and then go. (Alana, Egyptian Potential RCC Student Interview, May 11, 2021)

After a follow up question asking if she would attend ERU, Alana chuckled, stating that there were better options in Egypt, mentioning the global university rankings and reputation of American University in Cairo.

Scholarships to Russian Universities

As discussed previously, the former Soviet Union championed higher education development in Africa by providing millions of dollars in scholarships for African young people to study abroad in Russian universities. From 1955 to 1984, the most scholarships were awarded to students from Ethiopia, followed by Nigeria, Madagascar, the Republic of Congo, Ghana, and Tanzania. Although the number and amount of scholarship funding decreased dramatically near the end of this era, more money has become available again in the 2000s, along with rekindling diplomatic relationships with former partners on the continent (Holt et al., 2014).

As of 2021 there are close to 21,000 African students studying abroad in Russia, most of which come from Morocco, followed by Nigeria, Cameroon,

Zambia, and Kenya. Of all top five countries, Zambia currently receives the largest percentage of scholarships, and has a total of 6,000 students enrolled in Russian universities ("Number of students from Africa enrolled in Russian education institutions," 2020). However only a fraction of these students receive scholarships, with less than 2,000 scholarships awarded to Africa in 2020, due to the COVID-19 pandemic (Klomegah, 2021). Despite the positive impact of scholarship aid benefiting young Africans while simultaneously fulfilling the soft power agenda, there have been notable conflicts impacting the African students in Russia.

Examining popular Zambia newspapers, a quick search of the word "Russia" brought forth a large volume of articles about abuses Zambian students as well as other African students experience while studying abroad in Russia. This is even before adding the word "education" to the search terms. Some complaints include poor housing facilities and lack of amenities, while others allege that their scholarship living stipends have been delayed on multiple occasions by the Russian government, making it impossible to pay for everyday needs. A number of students, including the Zambian Student Union of Russia and Ukraine (ZASURU) contacted their government begging for aid to purchase food, leading to a public outcry. An excerpt from the published letter by ZASURU reads:

> This has been an ongoing issue for many years now. Students find themselves in very desperate situations due to delayed payments. This leads to students not being able to meet their obligations in terms of visa renewals, settling hostel fees, buying medical policies on time which in some cases have led to court summons for visa related issues and evictions from hostels for delayed payments. (ZASURU, 2016)

In another article from the year prior, Algerian, Moroccan, and Zambian students all experienced delayed stipend payments ("Zambian and Moroccan Students," 2015). It is interesting that this article was published in 2015, during the height of economic sanctions levied against Russia by the West. The ruble was low, and the country experienced economic hardship; one could infer those African students paid the price for what was not even their battle.

From Congo to GONGO: Russian Cultural Centers in Africa

In addition to fostering symbolic partnerships, establishing hybrid universities, and awarding scholarships, Russian Cultural Centers serve as a new type of institution within the current agenda of Russian higher education development in Africa. The prevalence of GONGOs offer new modes of learning around the world. A government-sponsored non-governmental organization (GONGO) is an institution type which straddles a fine line between NGO and Nation State status (Naím, 2007). Regimes utilize non-coercive strategies to consolidate power, avoiding international criticism on their nondemocratic domestic practices; GONGOs provide a discreet method for which governments influence and control NGOs.

Often considered citizen interest groups who lobby government actors, GONGOs have been a source of controversy in foreign policy since the 1980s (Lushnikov, 2019). For example, the National Endowment for Democracy (NED), is a GONGO funded and operated by the United States. Established in 1983 and partially funded by an allocation from Congress, the NED is marketed as a grant-making foundation that funds democratic initiatives abroad (ned.org). Despite the benign text of the mission, the NED has come under criticism many times over the years, due to a lack of transparency, as well as accusations of tampering with foreign politics (Anderson, 2004; Carothers, 1995; Corn, 1995; Drezner, 2000; Heymann, 1960; Nichols, 1990; Sims, 1990), with Naím (2007) nicknaming the NED a "neocolonial slush fund" (Naím, 2007).

Not only a U.S. and Russian strategy—countries throughout the world have their own GONGOs with a variety of political interests. For example, in China, the 1998 reforms to the State Council system caused a boom of environmentally focused GONGOs. The impact of these organizations expanded to work in collaboration with German NGOs on renewable energy and energy efficiency. The early founding of environmental GONGOs was a reaction to the "internationalization of environmental protection" in the late 1970s (p. 6), with China and other countries addressing issues together that continue to concern scientists worldwide (Wu, 2003).

There are two different types of cultural centers, founded in the late 2000s, that are funded and operated by various government-affiliated stakeholders in Russia. In total, between the two institutions, there are fifteen RCCs located across nine countries in Africa: Egypt, Ethiopia, Democratic Republic of Congo, Kenya, Morocco, South Africa, Tanzania, Tunisia, and Zambia (Popovic et al., 2020). The first type of institution—Russian Centers of Science of Culture, were founded by the Russkiy Mir Foundation, a GONGO established by a presidential decree. In his 2007 address to the Federal Assembly, President Putin stated:

> The Russian language not only preserves an entire layer of truly global achievements...As the common heritage of many peoples, the Russian language will never become the language of hatred or enmity, xenophobia or isolationism...In my view, we need to support the initiative put forward by Russian linguists to create a National Russian Language Foundation, the main aim of which will be to develop the Russian language at home, support Russian language study programs abroad and generally promote Russian language and literature around the world. (Putin, 2007)

A joint venture between the Ministry of Education and the Ministry of Foreign Affairs, Russkiy Mir is led by an academic dean and political figure, and the board of trustees consists of "prominent Russian academics, cultural figures, and distinguished civil servants." By nature, the organization is a tool of soft power, and subsists on both public and private funds. There are 109 RCCs established around the world via funds from the Russkiy Mir Foundation (Pieper, 2020).

The second type of RCC are the Russian Centers and Cabinets funded and operated by Rossotrudnichestvo, an autonomous agency funded and managed solely by the Ministry of Foreign Affairs. The main differences are that the first type, organized by Russkiy Mir are a collaborative effort between Foreign Affairs and Education, with funding from private donors, with the express mission of Language Education & Cultural Exchange. Whereas the second type organized by Rossotrudnichestvo, reports only to Foreign Affairs, and has a dual mission of Education & Foreign Aid. The impact of Rossotrudnichestvo has been met with controversy since its inception, with many considering the organization to be a recruitment tool, grooming foreign operatives for political use (Jurevičius, 2014; Popovic et al., 2020).

In contrast to the RCCs in the U.S. and in Europe, the RCCs in Africa seem to be the result of restructuring prior cultural centers, rather than building new relationships with universities. While the RCCs in Europe, the U.S. and Asia are often affiliated with local universities, there are less of these connections in Africa, even when there are local institutions available nearby. In Africa, RCCs offer a more comprehensive educational experience besides cultural engagement, with curriculum depending upon the need of the local community.

The RCC in Addis Ababa, Ethiopia produces more social functions such as networking events and business opportunities. The center website advertises alumni networking events for students who studied in the USSR/Russia and have returned to Ethiopia. The RCC Addis Ababa also helps students obtain recommendation letters to study in Russia, and links scholars who translate Russian literature into Amharic. In contrast, the RCC in Cairo, Egypt is strictly an educational facility, focused more on skills development, rather than sociocultural opportunities. The Cairo RCC could be considered vocational, due to the offering of a wide range of engineering, computer science and technology training courses taught in English.

Closing Thoughts

In considering the long impact of Eurasia on higher education development in Africa, one must connect the political goals with the education goals of each nation state. Savage (2016) relates education and politics in this way, stating (p. 42):

> It was often the students themselves who exposed the frictions between the various registers of internationalism.... A close reading of these histories therefore demands a disentangling of multiple narratives of political desire, and an excavation of solidarity at the micro-level, in the lived experiences of the people who undertook these journeys. (Savage, 2016)

I offer an outsider's examination of new modes of learning in informal higher education in Africa, while tracing the political goals and impact of Russian educational engagement on the continent. This research inquiry offers a three-pronged approach to study this problem: examination of the national goals that catalyze the development of these programs, investigation of the institutional practices and purposes of cultural centers, and analysis of individual student motivations and experiences with these educational institutions.

References

Abdelmoaty, G. A., & Soliman, S. A. E. M. (2020). Smart Technology Applications in Tourism and Hospitality Industry of The New Administrative Capital, Egypt. *Journal of Association of Arab Universities for Tourism and Hospitality*, 19(2), 102–129.

Anderson, R. D. (2020). German (Humboldtian) University Tradition, The. *The International Encyclopedia of Higher Education Systems and Institutions*, 546–551.

Arts, M. o. T. a. *Traditional Ceremonies*. Retrieved August 31 from https://www.mota.gov.zm/?page_id=5407

Arts, M. o. T. a. (2020). *Traditional Ceremonies*. Retrieved August 31 from https://www.mota.gov.zm/?page_id=5407

Ayembe, D. (2021). *Egypt's New Administrative Capital Project Timeline and What You Need to Know*. Construction Review Online. Retrieved July 4 from https://constructionreviewonline.com/project-timelines/egypts-new-administrative-capital-project-timeline-and-what-you-need-to-know/

Blumer, H. (1986). *Symbolic Interactionism: Perspective and Method*. University of California Press.

Bourdieu, P. (1979). Symbolic Power. *Critique of Anthropology*, 4(13–14), 77–85.

Bourdieu, P. (2011). The Forms of Capital. (1986). *Cultural Theory: An Anthology*, 1, 81–93.

Chankseliani, M. (2021). The Politics of Exporting Higher Education: Russian University Branch Campuses in the "Near Abroad". *Post-Soviet Affairs*, 37(1), 26–44. https://doi.org/10.1080/1060586x.2020.1789938

Clarke, J. C. (2011). *Alliance of the Colored Peoples: Ethiopia and Japan before World War II*. Boydell & Brewer Ltd.

Coles, R. (1999). Pushkin's Black Consciousness. *CLA Journal*, 43(1), 54–72.

Cochran, J. (2012). *Education in Egypt (RLE Egypt)* (Vol. 1). Routledge.

Contributor. (2018). We Plan to Create an Ethiopian Center for Nuclear Science and Technologies. *The Reporter*. https://www.thereporterethiopia.com/article/we-plan-create-ethiopian-center-nuclear-science-and-technologies

Demassie, A. A. (2021). *Global Perspectives: Ethiopia-Russia Relations* [Interview]. The Kennan Institute. https://www.wilsoncenter.org/event/global-perspectives-ethiopia-russia-relations

Drake, S. C. (1987). *Black Folks Here and There*. Los Angeles: University of California, Los Angeles Press.

Egypt, Russia to Cooperate in ICT Sector, Africa. (2019). https://www.egypttoday.com/Article/1/67263/Egypt-Russia-to-cooperate-in-ICT-sector-Africa

Egyptian Russian University. (2021). Retrieved July 4, 2021 from https://www.eru.edu.eg/

Farberman H. A. & Perinbanayagam R. S. (1985). *Foundations of interpretive sociology : original essays in symbolic interaction*. JAI Press.

Fikade, B. (2019). Russia Agrees to Provide Atomic Energy Supplies to Ethiopia *The Reporter*. https://www.thereporterethiopia.com/article/russia-agrees-provide-atomic-energy-supplies-ethiopia

Foglesong, D. S. (2014). *America's Secret War Against Bolshevism: US Intervention in the Russian Civil War, 1917–1920*. UNC Press Books.

Gaddis, J. L. (2006). *The Cold War: A New History*. Penguin.

Gerőcs, T. (2019). The Transformation of African–Russian Economic Relations in the Multipolar World-System. *Review of African Political Economy*, 46(160), 317–335. https://doi.org/10.1080/03056244.2019.1635442

Hartmann, S. (2008). "At School We Don't Pay Attention Anyway" – The Informal Market of Education in Egypt and Its Implications. *Sociologus*, 58(1), 27–48. http://www.jstor.org/stable/43645616

Holt, J., Newhouse, S., & Ukhova, D. (2014). *Scholarships and the Healthcare Human Resources Crisis: A Case Study of Soviet and Russian Scholarships for Medical Students from Ghana*. Oxfam GB.

Jurevičius, M. A. (2014). "Russkiy Mir" Concept–Russia's Strategic Centre of Gravity. *Ad Securitatem, 121*.

Klomegah, K. K. (2021). *Fewer African Students Came to Russia in 2020* https://www.moderngh ana.com/news/1072427/fewer-african-students-came-to-russia-in-2020.html

Kochetkova, I. (2009). *The Myth of the Russian Intelligentsia: Old Intellectuals in the New Russia* (Vol. 62). Routledge.

Kotek, J. (2015). *Students and the Cold War*. Springer.

Kuraev, A. (2014). *Internationalization of Higher Education in Russia: Collapse or Perpetuation of the Soviet System? A Historical and Conceptual Study* [Dissertation]. Boston College.

Lilge, F. (1968). Lenin and the Politics of Education. *Slavic Review*, 27(2), 230– 257.

Loewert, P., & Steiner, C. (2019). The New Administrative Capital in Egypt: The Political Economy of the Production of Urban Spaces in Cairo. *Middle East- Topics & Arguments*, 12, 66–75.

Lushnikov, D. A. (2019). Government-sponsored non-governmental organizations (GONGO): genesis of the problems, interpretation and functions. *Polis. Political studies*, 2(2), 137–148.

Martin, T. D. (2001). *The Affirmative Action Empire: Nations and Nationalism in the Soviet Union, 1923–1939*. Cornell University Press.

Martone, E. (2008). *Encyclopedia of Blacks in European History and Culture [2 volumes]*. ABC-CLIO.

Mathews, S. (2021). The Competition for Egypt: China, the West, and Megaprojects. *Al Jazeera*. https://www.aljazeera.com/news/2021/3/15/the-competition-for-egypt-china-the-west-an d-mega-projects

Matusevich, M. (2008). An Exotic Subversive: Africa, Africans and the Soviet Everyday. *Race & Class*, 49(4), 57–81.

Mcclellan, W. (1993). Africans and Black Americans in the Comintern Schools, 1925–1934. *The International Journal of African Historical Studies*, 26(2), 371. https://doi.org/10.2307/219551

Mohamed, R. Y., & Trines, S. (2019). *Education in Egypt* (Education System Profiles, Issue. https://wenr.wes.org/2019/02/education-in-egypt-2

Mortimer, R. A. (1972). Soviet-African Relations-Soviet Policy in West Africa. Robert Legvold. (Cambridge, Massachusetts: Harvard University Press, 1970, 372 pp. $13.00). *African Studies Review*, 15(2), 347–349.

Nash, M., Cowcher, K., De Oliveira, A. B., Ribeiro Sanches, M., Savage, P., Siegert, N., Vasić-Janeković, V. (2016). *Red Africa: Affective Communities and Cold War*. Black Dog Publishers.

Naím, M. (2007). What Is a GONGO? *Foreign Policy*, 160, 96.

Number of Students from Africa Enrolled in Russian Education Institutions. (2020). In: Statista Research Department.

Patman, R. G. (2009). *The Soviet Union in the Horn of Africa: The Diplomacy of intervention and Disengagement* (Vol. 71). Cambridge University Press.

Pham, J. P. (2010). Back to Africa: Russia's New African Engagement. In *Africa and the New World Era* (p. 71–83). Springer.

Pieper, M. (2020). Russkiy Mir: The Geopolitics of Russian Compatriots Abroad. *Geopolitics*, 25(3), 756–779.

Popovic, M., Jenne, E. K., & Medzihorsky, J. (2020). Charm Offensive or Offensive Charm? An Analysis of Russian and Chinese Cultural Institutes Abroad. *Europe-Asia Studies*, 72(9), 1445–1467. https://doi.org/10.1080/09668136.2020.1785397

Putin, Vladimir. (2007). *Annual Address to the Federal Assembly*. Retrieved January 29, 2023 from http://en.kremlin.ru/events/president/transcripts/24203

Rezaev, A. V. (2006). Diversification in Russian Higher Education: Profiles, Foundations and Outlooks. In *Higher Education in a Global Society: Achieving Diversity, Equity and Excellence* (pp. 107–124). https://doi.org/doi:10.1016/S1479-358X(05)05005-9

Richards, A. (1992). *Higher Education in Egypt* (Vol. 862). Washington, DC: World Bank.

Rosen, S. M. (1963). *Soviet Training Programs for Africa* (Vol. 9) [Bulletin]. US Dept. of Health, Education, and Welfare, Office of Education.

Russell, M., & Pichon, E. (2019). Russia in Africa. A New Arena for Geopolitical Competition. *Briefing, European Parliamentary Research Service*, 8.

Saunders, F. S. (2013). *The Cultural Cold War: The CIA and the World of Arts and Letters*. The New Press.

Savage, Polly (2016) 'Reading Between the Lines: African Students in the USSR.' In: Nash, Mark, (ed.), *Red Africa: Affective Communities and the Cold War*. London: Black Dog Publishing, pp. 35–43.

Schmemann, S. (2010). Of African Princes and Russian Poets. *The New York Times*, 12.

Schmidt, E. (2013). *Foreign Intervention in Africa: From the Cold War to the War on Terror*. Cambridge University Press.

Shumba, A. (2016, June 14, 2016). Chabuka Ceremony. *Music in Africa*. https://www.musicinafrica.net/directory/chabuka-ceremony

Smolentseva, A. (2003). Challenges to the Russian academic profession. *Higher Education*, 45(4), 391–424.

Tareke, G. (2009). *The Ethiopian Revolution: War in the Horn of Africa*. Yale University Press.

The Delegation of Russian Universities in Ethiopia. (2019). Russian Center for Science and Culture Addis Ababa. Retrieved June 26, 2021, from https://eth.rs.gov.ru/en/news/61025

The Raven Baron, The Photo Album and A Lost World Recovered: Our Brother Fitawrari Tekle Hawariat of Ethiopia. (2021). Retrieved July 4 from https://kotchoubey.com/select-biographies/petia-the-abyssinian-our-brother-tekle-hawariat-of-ethiopia/

Tula, L. S. (2018). Chabuka Ceremony (J. I. Bwanga, Trans.). In @zambiatraditionalceremonies (Ed.): Facebook.

Van Creveld, M. L. (1990). *The Training of Officers: From Military Professionalism to Irrelevance.* New York, NY: Free Press.

Weaver, H. (1985). *Soviet Training and Research Programs for Africa.*

Wu, F. (2003). Environmental GONGO Autonomy: Unintended Consequences of State Strategies in China. *The Good Society, 12*(1), 35–45. https://doi.org/10.1353/gso.2003.0031

Yakobson, S. (1963). The Soviet Union and Ethiopia: A Case of Traditional Behavior. *The Review of Politics, 25*(3), 329–342.

ZASURU. (2016). *Zambian Students in Russia, Ukraine Not Given Allowances.* https://www.zambiawatchdog.com/zambian-students-in-russia-ukraine-not-given-allowances/

· 3 ·

RESEARCH CONSIDERATIONS AND THEORETICAL APPLICATIONS

Abstract: This chapter discusses the role of government-sponsored non-governmental organizations (GONGOs) in educational initiatives in Africa. Using Russian Cultural Centers in three African countries as cases, this chapter teases out how sociocultural values and practices influence political partnerships.

This chapter describes the research methodology and offers the major findings of this work by focusing on the historical and political context behind the development of this new type of institution, and the personal motivations that guide African students' engagement within them. The analysis of the data is framed through two perspectives: the State and the Student. The State includes the USSR, the Russian Federation, and the governments of Egypt, Ethiopia, and Zambia. Although these diverse state actors have different goals and historical contexts, they all wield political power over the Student.

The countries were chosen to represent geographic regions and historical sociopolitical contexts: Indigenous afrosocialist (Zambia) vs. Marxist-Leninist (Ethiopia) vs. Arab socialist (Egypt). This research relies on qualitative methodology including digital archival research, content analysis, a focus group, one on one interviews, and observation of the Zambian RCC's debate team competition. Data was triangulated through multiple methods and data sources to increase the credibility of findings.

Keywords: GONGO, Cultural Centers, History, Socialist Identity, Qualitative Research Methodology, Egypt, Ethiopia, Zambia, Diplomacy Geopolitics, African Education

Introduction

Russian Cultural Centers (RCCs) are an example of the new types of cross-national partnerships developing in higher education in Africa. This chapter offers the major research findings of this work by focusing on the historical and political context behind the development of this new type of institution, and the personal motivations that guide African students' engagement with these institutions. The analysis of the data is framed through two perspectives: the State and the Student. The State includes the USSR, the Russian Federation, and the governments of Egypt, Ethiopia, and Zambia. Although these diverse state actors have different goals and historical contexts, they all wield political power over the Student. Here, the "Student" is defined as: alumni of Soviet-era educational programming, current students enrolled in RCCs in Africa, and the general African public, which includes potential RCC students.

I consider the State as political actors who manipulate the student (future worker) to fulfill national needs, but I do not differentiate between Russia and Africa agendas, challenging the outdated notions of spheres of influence. As governments align based on values and goals, their needs align, sometimes at the expense of their own peoples. I frame this analysis by highlighting the power differential between the two stakeholders, regardless of GDP, military might, and geopolitical power.

In this chapter I begin with a summary of the higher education development that occurred during the era of the Soviet Union and the current Russian Federation. Then, I discuss the two major themes that arose from the findings of this research: (1) the development of the African Elite, and (2) dignity, political camaraderie, and respect. The chapter closes with the research methodology and theories that inform this discussion.

Higher Education Development in Africa during the Era of the Soviet Union

After the establishment of the USSR in 1922, the purpose of higher education in Eurasia shifted from educating only the Russian elite, expanding to educate all people. Marx inspired a romantic and utopian vision of an educated populace, whereas the proletariats implemented a practical vision of the people—educated for the purpose of serving as workers. Lenin combined these two purposes, with a vision of disseminating socialist ideology and lifting the oppressed from the mental bondages of their colonial oppressors, while simultaneously building infrastructure and training a new generation of workers (Kuraev, 2014).

With internationalism a vital facet of his political and social activist agenda, Lenin viewed the development of higher education in Africa as an opportunity to build political relationships and train new allies. One of the ways the USSR was involved in higher education development in Africa was in the support of like-minded political goals in newly liberated nations. A poignant example of this goal is the now defunct Stalin Communist University of the Toilers of the East (KUTV) in Moscow. When African students began studying at KUTV the in the 1920s, their coursework included language study, critical theory, and the sciences. But there was also a hidden curriculum training students in espionage, with courses in small arms training, coding, guerilla warfare, rule of conduct under surveillance, and interrogation techniques (Nash et al., 2016).

After the death of the decidedly xenophobic party leader Joseph Stalin, there was a surge of international engagement. Khrushchev (the new premier after the short-lived Malenkov), sought political friendships with African leaders in exchange for technical education to increase the numbers of pilots, medical experts, and construction workers in their countries (Nash et al., 2016). The scope of Soviet programming from the 1950s onwards included the establishment of Patricia Lumumba University (now RUDN University) to serve foreign students, technical training programs in Africa, scholarships to attend Russian universities, and short-term noncredit cultural exchange programs targeting athletes, musicians, and future students (Filatova, 2001; Nash et al., 2016; Weaver, 1985).

However, despite the egalitarian ideals and positive intentions, there were conflicts and outright hostilities that contradicted these partnerships. Racism and xenophobia combined with the resource scarcity of the Gorbachev years lead to an exodus of foreign students no longer welcome in the waning years of a great superpower (Nash et al., 2016). Despite the termination of these programs upon collapse of the USSR in 1991, the legacy of Soviet soft power is a positive one, considering the vast impact on the economy and infrastructure of the African continent (Kotek, 2015).

Higher Education Development in Africa during the Era of the Russian Federation

After the political transition from USSR to the Russian Federation, President Yeltsin's era is characterized by missed opportunities to engage with Africa. During this time Russia shut down many of their embassies, consulates, and cultural centers on the continent, downsizing their soft power influence (Schmidt, 2013). However, one might argue Yeltsin' hands were full managing the economic hardship, food scarcity, and internal conflicts in his own country. In the declassified meeting minutes from a 1995 conversation between German Chancellor Kohl and American President Clinton, Kohl alleged that Yeltsin was a decorative figurehead unable to control his own troops:

> I am not one who prays to icons in the corner. I don't know if Yeltsin will prevail. But I am sure that if we leave him in the lurch, matters will get much worse. . . . We need to talk with Yeltsin. I don't like calling him every week and spelling out for him how Russia's image is going downhill. . . .I am not sure whether it is malice or ineptness or both on the part of the Russian military, but he has a military who cheated him. They set him up for a situation in which he could only fail. They set a trap for Yeltsin. . .perhaps to topple him. (Council et al., 1995)

After former Prime Minister (and sixteen-year KGB agent) Vladimir Putin was elected president in 1999, he enacted a series of economic reforms, raising the country's GDP by 72 %, leading to a decline in poverty and unemployment throughout the nation. After resolving these pressing internal priorities, Putin then began a robust foreign policy agenda to revive the image of the Russian Federation (Åslund et al., 2010). Following his 2007 public announcement of establishing RCCs around the world, Putin stepped down into a Prime Minister role from 2008–2012 (before returning to the presidency from 2012 to the present). It was during this time that then-President Medvedev visited four African countries for official visits: Angola, Egypt, Namibia, and Nigeria. This 2009 trip has been described as the official revival of Russia's interest in Africa (Filatova, 2001; Russell & Pichon, 2019).

Now, the Russian Federation has resumed its policy of awarding scholarships to African students, as well as engaging in symbolic partnerships, building hybrid universities, and establishing Russian Cultural Centers. As discussed in Chapter 2, the USSR valued the study and preservation of indigenous African languages. Now it is the reverse, with RCC's expressed goal of training Africans to speak Russian and attend Russian universities. The findings of this study argue that the result of the legacy of Eurasian higher education development in Africa is the development of the African Elite, and the cultivation of political camaraderie and mutual respect.

Development of the African Elite

> Let us agree to call African elitism the entitlement to an uncontested leadership inferred from the privilege of being exposed to modern education...It is as though Westernization passes on to local elites the right to rule; that is, to continue the unfinished business of colonialism. (Kebede, 2003, pp. 167–168)

In East and South Asia, a Portuguese term is used: "comprador", originally referring to indigenous servants in the homes of European colonists. The term has now evolved, referring to locals who intercede for foreign organizations, whom are involved in political and economic and exploitation of their own people (Po-Keung & Tak-Wing, 1999). The Belgians and the French used the term "évolués" to refer to the indigenous Africans who left behind their cultural traditions in favor of those of their colonizers (Kebede, 2003). Also called "les noirs perfectionnés" (the perfected blacks), évolués held white collar jobs, enjoyed power and privilege in society, and often acted as mediators between the colonists and their own kin, who had not been elevated to this status (Tödt, 2012).

Much has been written on the vacuum left by colonizers post African liberation, the subsequent appointment of "perfected blacks," and the chaos and corruption that followed, destabilizing regions already at the brink of collapse (Gyekye, 2015; Hyslop, 2005; Lange, 2004; Ojo, 2018; Pierce, 2006). Although the Soviet Union never colonized Africa, and Ethiopia never fully succumbed to European occupation, the impact of Soviet higher education engagement in Africa led to the development of an elite class of African allies. This practice of uplifting and supporting the Soviet-educated elite into positions of power is a prime example of neocolonial behaviors exhibited by countries without a past history of colonization.

In Egypt

Former Egyptian president Hosni Mubarak attended a Soviet pilot training school in Kyrgyzstan from 1959–1961, and again for advanced studies in Moscow in 1964, before his career as an officer in the Egyptian Air Force. Mubarak's education occurred during the years of President Gamal Nasser, who's anti-imperialist policies and support of Arab socialist ideals led to friendly relations between Egypt and the USSR. These relations were severed when Egyptian President Anwar Sadat came into office, when he conspired with President Nixon to dissolve his relationship with the Soviet Union. In 1981, when Anwar Sadat was assassinated (while seated right next to Mubarak in a car). Mubarak, who happened to be vice president at the time, ascended to power. Immediately after taking office Mubarak rekindled relations with the USSR, and by 1984, appointed an Ambassador from Egypt to Russia, exactly twenty years after his time as a pilot cadet in Moscow (Arafat, 2011; Shehata, 2011).

However, a Soviet education does not necessarily mean a lifetime of alliance. After thirty years in power as president of Egypt, Mubarak was long-time considered an ally of the West. One example being when Mubarak established membership in the allied coalition in the Gulf War, when Egyptian boots were first on the ground in Iraq, later rewarded by the United States who led the efforts to forgive twenty billion dollars of Egyptian debt (Al-Awadi, 2005). This savvy maneuvering is an example of the critical geopolitics that frame this work. Although the tyranny of Mubarak and his support of neoliberal agendas is at odds with the critical theory aspect of subaltern geopolitics, his policies strategizing in favor of his home country (a former colony) support this frame. The USSR and now the Russian Federation may have utilized soft power in Africa to expand their influence, but African leaders shift their alliances as they see fit, subverting the idea of superpowers acting on "weaker" countries. Through critical geopolitical maneuvering, Mubarak went from being the Student, to wielding the power of the State.

In Zambia

> The reason why we are saying it would be better [to study there] is because one of our ministers actually studied in Russia. She's a minister now. So, if we have a lot [of Zambian students] going there, imagine having a lot of *hers* in the world, in the whole country. [All laugh] So meaning our country would be somewhere, big ideas would be building. (Lucy, Focus Group with Zambian RCC on April 30, 2021)

In the quote above, Lucy, a nineteen-year-old RCC student from Lusaka, Zambia, speaks about Zambian politician: Dr. Nkandu Phoebe Luo, during a focus group with other students from the RCC in Lusaka. Dr. Luo is a scientist who received her master's degree from Moscow State University in 1977, before attending medical school in Brunei (*Profile of Nkandu Luo*, 2021). Moscow State is considered the Harvard of Russia and the country's most prestigious of institutions, respected throughout Eurasia. Both a microbiologist and a professor at the University of Zambia, for Lucy, Dr. Luo represents the epitome of Zambian achievement, and evidence of the potential of a Russian education abroad. A success story, yes, but there is also a deeper analysis, here.

Dr. Luo's research background and advocacy work on HIV/AIDS led to her appointment as Minister of Health (1997–1999), followed by an appointment as Minister of Transportation (1999–2001), then Minister of Local Government and Housing (2011–2014), then Minister of Chiefs and Traditional Affairs (2014–2015), Minister of Gender (2015), Minister of Higher Education (2016–2019) and finally her current role as Minister of Fisheries and Livestock (2019–2021) (*Profile of Nkandu Luo*). The constant movement from office to office could be due to her value to the government, shifting her into various roles that need her expertise. Or, this movement is the result of internal conflicts influenced by external solutions.

In December of 1999 during Dr. Luo's first political role as Minister of Health, a group of nurses and 300 doctors went on strike in protest of extreme work conditions and lack of funding. In response, Dr. Luo fired all 300 doctors and replaced them with doctors flown in from Cuba, who received higher pay and benefits. This led to an exodus of health professionals from Zambia in the early 2000s, many moving to other African countries or the United Kingdom (Makasa, 2008; Mukwita, 1998; Schatz, 2008; *The disaster of Nkandu Luo as minister of health in 1999*, 2021). It was Dr. Luo's transnational social capital developed as a result of her education in the USSR, that brokered the recruitment of Cuba doctors that replaced her Zambian colleagues and countrymen.

As discussed in Chapter 2, Cuba and the Soviet Union worked hand in hand in support of the liberation of colonized nations in both Africa and Latin America. Fidel Castro sent 300,000 troops to engage in the Kremlin's proxy wars during the Cold War era (Gleijeses, 2002). In describing the USSR and Cuba's military influence in Angola in the late 1970s, former Zambian president Kenneth Kaunda once stated "Africa has fought and driven out the ravenous wolves of colonialism: racism and fascism out the front door, but a plundering tiger, with its deadly cubs, is now coming through the back door" (Small, 1977). In describing Cuba as deadly cubs, the children of plundering tiger the Soviet Union, Kaunda identifies a familial relationship between the two nations, united in military customs and ideological goals. It is no surprise that Luo, who was educated in the USSR at the height of friendship between their three governments, would maintain ties with their allies (and her colleagues and friends from Moscow State University), even at the expense of her own people.

If not for her Soviet education and the social ties she created during graduate school (with fellow doctors from Cuba), Dr. Nkandu Luo would never have reached her career accomplishments, which led to her government appointments, and her ability to wield power over the livelihoods of hundreds of medical professionals. I also offer this vignette as justification for combining both African governments and Russia in my analysis of the Perspectives of the State. In this instance, Dr. Luo's actions in 1999 had no political benefit for Russia, but they did for the Zambian government. Luo's status is an example of the neocolonial intentions that manifest in the development of an elite class of Africans.

In 1968, former CIA director Richard Bissell connected the necessity of developing and manipulating local actors, when he described the steps of covert operations, beginning with finding allies within the country for whom there would be mutual benefits to cooperating with foreign powers (Marchetti et al., 1974). Berman (1974) discusses clientelism, which is the exchange of political favors for goods and services, arguing that as the government was often the largest employer in African countries during the Cold War—the marker of elite or pre-elite status, is civil service and public employment, creating an elite class of senior officials and administrators, and their families and friends (Berman, 1974).

Another topic impacting Zambian students is the tense matter of *who* is awarded scholarships to study abroad in Russia. According to the students from the RCC in Zambia, 80 % of their students who apply for scholarships to Russia, receive them. However, there is no requirement to attend an RCC to win a scholarship. In fact, the data points to a trend of funding only the wealthy elite such as the children of Zambian being political leaders. In the previously cited article about late stipends to Zambian and Moroccan students, the unnamed reporter took a jab at Zambian Minister of Agriculture: Given Lubinda. The article reposted a personal photograph of the politician on a recent trip to Russia to visit his daughter Namatama, who was at that time a college student in Moscow. The final sentence of the article reads:

> Lubinda was last week in Moscow to take food and money to his daughter at one of the universities. We just hope Lubinda's daughter will share with her friends who are starving. ("Zambian and Moroccan Students," 2015)

This pointed statement earned the wrath of Namatama Lubinda herself, who published an op-ed response to the article. Lubinda defends her father stating that he paid for her siblings to study in China and Malaysia, and also offered to also cover the cost for her to study in China. However, she applied to a Russian scholarship without his knowing, as a gesture of respect since he already supports her siblings. She states that she received the scholarship based on her grades, implying that this was not a political maneuver (Lubinda, 2015). However, in this op-ed Lubinda embodies the old saying: "not all skin folk are my kinfolk" when she does not hesitate to deny the experiences of her classmates, even denigrating them in the process. One excerpt reads:

> First of all, he is not responsible for the children on scholarship. That's the responsibility of the ministry of education...Secondly, you portrayed a picture that I do not deserve the scholarship that I am on, sponsored by the RUSSIAN GOVERNMENT. Please understand the meaning of the word "scholarship". According to the oxford dictionary...the word "scholarship" means "A grant or payment made to support a student's education, awarded on the basis of academic or other achievement" Do you see the word "poverty" in that? It is awarded on the basis of ACADEMIC ACHIEVEMENT which you, and many other Zambians don't seem to understand that I worked hard to achieve. I graduated my high school with STRAIGHT DISTINCTIONS and therefore, the bursaries committee saw the need to fairly grant me the scholarship.

Another fiery passage from Lubinda's op-ed states:

> Concerning our allowances. My fellow students, please don't accuse the bursaries committee of starving us for 3 months. According to our contracts, we are entitled to $1000 every 4 months...Be responsible with your money!!! Of course they delay but don't exaggerate everything especially seeing the current situation of our country. Anyway, because we all haven't been paid and people are starving should not be my problem or my father's problem. Please don't involve is in things that don't concern us.

In the closing of her piece, Lubinda addresses the newspaper directly, stating:

> My advice to you is that you provide the country with concrete and helpful information and not petty and private issues which don't concern the country at large. As a public media page, I suppose you are supposed to be a lot more responsible than that and not do "Chinese whisper" because that's not professional. (Lubinda, 2015)

The elitism is apparent in the quotes above, not in the touting of academic achievements, but in the othering of her fellow citizens. It could be that her father is unjustly targeted, as he has a rocky background in politics, and his decision to send all of his children abroad to study rather than at the University of Zambia, is in line with the practices of the upper class in many developing countries. Economic advancement and improved infrastructure are not the only results of an educated populace; the impact of Soviet and Russian higher education engagement also led to the development of an elite class of allies in Zambia.

In Ethiopia

As a postsecondary degree is required for many of these positions, the educated, wield both intellectual and political power, often with transnational influence. Back in 1979, Sklar identified the power inherent in education, and its linkages to the elite and their subsequent government power, stating:

> Ethiopia has a modern ruling class...Nurtured in urban centers and small towns, the core of this class is an educated administrative elite, recruited from the families of landowners, merchants, and officials...Prior to the revolution of 1974 and the subsequent abolition of private property in land, members of this class were closely associated with landlordism...The revolution has now blocked this avenue of enrichment, but the main forces of class formation—modern education and public employment—are likely to sustain the vitality of Ethiopia's ruling class whatever may become of the practice of landownership (Sklar, 1979, pp. 534–535).

In this quote Sklar posits that modern education and public employment were the major forces of class formation in Ethiopia. Similarly, Vladimir Lenin stated that with education and electricity, Russia would eventually reach the final stage of development: communism (Lenin, 1920). The concept of an elite class is not only a Western notion linked to colonialism. In 1902 in an infamous pamphlet, Vladimir Lenin wrote the bones of his ideological vision. In this publication, he sets up the seeds of Leninism, in response to the writings of Marx. Many consider this work to be elitist, as it advocates for a centralized government moderated by a group of professional revolutionaries (Gleberzon, 1978). It is in these writings that Lenin develops his concept of the intellectual, stating that in order to overthrow the aristocratic bourgeoisie, *all* workers must become intellectuals (Lenin, 2012).

One must remember that Nadezhda, Lenin's wife at this time, (and who's name translates to "Hope") was a schoolteacher, and vital in her husband's efforts to improve and expand education. Immediately after the October Revolution of 1917, Nadezhda became deputy commissioner of education, and later developed an adult education division. In Lenin's shadow, Nadezhda devoted considerable time and resources to the development of the educated Soviet worker, influencing her husband's political and ideological agenda (Krupskaya, 1940).

The concept of the intellectual within Leninism maintains that the everyday worker, including teachers, administrators, and scientists, should all aspire to be political leaders, stating that power should be placed in the hands of the educated (not those born into a particular bloodline) (Lilge, 1968)). Kebede (2003) connects the impact of Leninism to the development of the Ethiopian elite, pointing to the Ethiopian college student revolutions of the 1960s and 1970s, in opposition to the ancient feudal practices that had characterized the economy for centuries. The brief rule of the Derg (Ethiopian communist party), coupled with the Marxist education Ethiopians received in the USSR advanced this notion that politicians should be educated (Kebede, 2003).

Dignity, Political Camaraderie, and Respect

Another theme that arose from the data were the concepts of dignity, political camaraderie, and respect. Continuing on the topic of Ethiopia, we must consider the cultural connections between the USSR and Abyssinia (the former name of the Ethiopian kingdom and its feudal territories). Demassie (2021) articulates four guiding reasons for the longstanding relationship between the regions: (1) an ancient global presence with a multiethnic population, (2) the influence of Marxism, (3) long-term military cooperation, and (4) the influence of the Orthodox Christian church (Demassie, 2021). Both Marxism and Orthodoxy fall into the concept of value homophily, which guides the theoretical framework of this research.

The Ethiopian Orthodox Tewahedo Church is the largest of the six Oriental Orthodox Churches (which include Armenia, Egypt, Syria, India and Eritrea), and falls within the tradition of Eastern Christianity, which includes fourteen Eastern Orthodox Churches spread across Russia, Eastern and Southern Europe (Jenkins et al., 2018). Considering the historical geographic coverage of Eastern Christian churches compared to Western Christian churches (which include the Roman Catholic Church and the many protestant offshoots), there is a clear divide, expanding far beyond religion. Rupprecht (2018) on p. 215 encapsulates the political agendas woven into the East-West dichotomy of Christendom:

> Orthodoxy's role in modern global politics in fact went far beyond Cold War espionage. The entangled history of modern Russia and Ethiopia elucidates how Orthodoxy was a source of cross-border identity, legitimized political decision-making, and was a force for mobilizing and controlling populations. This was true even in states that repressed believers and considered expressions of spirituality to be backwards. (Rupprecht, 2018)

In an interview with an Ethiopian newspaper in 2018, Russian ambassador Tkachenko describes the origins of Russian-Ethiopian relations, first starting with religion, and then immediately transitioning into the topic of past military assistance. These two sources of human conflict: religion and war, highlight the complex weaving of culture (soft power) and military (coercive power), impacting the current geopolitical agenda.

> Coming back to the origins of our historical friendship, it all began with contacts between our orthodox churches in the middle of the 19th century. They laid the basis on which we started to build our further cooperation. In 1895 the Russian Empire provided the Abyssinian army with 30,000 rifles and 5,000 sabers to oppose the Italian aggression. Russian volunteers fought shoulder to shoulder with Ethiopian brothers in the glorious Battle of Adwa. Moreover, the Russian Red Cross mission arrived in Abyssinia to take care of the sick and wounded. (Tkachenko, 2018)

Demonstrating the shared values between the countries, Tkachenko states:

> I don't think there are many examples of such relations that are marked with true friendship and mutual trust throughout centuries. Common orthodox faith was the foundation that brought us closer together. Faithful Russians and Ethiopians celebrate same religious holidays, observe same fasts, and follow same orthodox traditions in family and social life. In fact, we are now intensifying our spiritual ties. Last month there was an exchange of religious delegations to Moscow and Addis Ababa and now we are expecting Patriarch Abune Mathias I to visit Russia for the first time. It will open a new page in the history of relations between the two sisterly churches. (Tkachenko, 2018)

Although many scholars may blame technology and imperialism for globalization, religion has always been the prime driving force behind the melting of borders in favor of a solidified ideological regions of thought. From the early Christian missionaries that laid a path for colonial soldiers, to the military expansion of the Muslim Ottoman Empire, religion not only justifies wars and expansion efforts, but also ties together a cultural and emotional dependence between its followers, and their country's governments. Indeed Marx and

Engels stated that "religion is the opium of the masses" describing the narcotic, blinding effect of mixing Church and State (Marx & Engels, 1964).

Religion also serves a form of transnational social capital, a prime example being the performance of priestly rites by Ethiopian orthodox priests in the USSR and Russia, and vice versa. The first African students to study in Russia were Ethiopian military cadets, long before the establishment of the Soviet Union. Their religious and therefore cultural proximity made these students the ideal guinea pigs in what would be a long history of educational partnerships. Therefore, membership in an orthodox church is example of both value homophily and transnational social capital, playing out in the enrollment of Ethiopian students in Russian educational training.

Two Ethiopian alumni of Soviet-era programming discuss sociocultural proximity and positive social engagement in the following two quotes:

> You have the understand the Russian they like us because we have a long history. The church, many are Orthodox, and in ancient times they travel to Ethiopia and meet the priests. Not me, I'm born again [evangelical Christian], many years. But the history, the culture, the Russians respect us more than others because of the [Orthodox] church. (Yonas, Ethiopian Alumni Interview, April 2, 2021)
>
> When I went to Leningrad, it was called Leningrad back then, it was 1978. It was a training program. We were [athletes] very serious. We learn with them for a couple months. Russian coaches and athletes with Ethiopians. The Russians they were great, they pay for everything, they treat us very well, like Kings! At the hotel, the girls would be in line, down the street, waiting for us, calling for us. (Abraham, Ethiopia Alumni Interview, March 14, 2021)

In the two quotes above, Ethiopian former professional athletes discuss their participation in short term cultural exchange programs that took place in 1978 in Saint Petersburg. Both highlight the social aspect of their experience in the USSR, using words like "respect" and being treated "like Kings." Although Abraham inadvertently touches on one of the main sources of sociocultural tension—young Russian women seeking African romantic trysts, triggering the rage and jealousy of their Russian male peers, often ending in violence towards African male students (Law, 2016; Leviyeva, 2005; Martone, 2008; Nash et al., 2016; Weaver, 1985).

In comparing the official social media pages of the various African RCCs, I noticed multiple online advertisements to apply for Russian universities. These ads would tout the scholarships available, and welcoming environment for African students. In some cases, instead of picturing African students engaged in learning, or the architecture of Russian universities—these ads featured photographs of young, blonde, pretty Russian women, depicted holding a diploma or book, while staring into the camera. It seems the romantic motivations of young African men are being marketed to, despite the potential for conflict.

Russian ambassador Sergey Lavrov also emphasized positive social engagement in a 2018 interview with Ethiopian newspaper *The Reporter*, stating:

> Russia values the long-established friendly relations with Ethiopia. I am pleased to note that our wide-ranging cooperation is built on the principles of equality, mutual trust and respect. Russia and Ethiopia maintain intensive political dialogue underpinned by the concurrence or considerable closeness of our views on the key problems of our time. (Contributor, 2018)

Minister Tkachenko may have received the same talking points that year, as he also stated in an interview:

> The victory of the Ethiopian resistance over Italians in 1941 gave the world the first glimmer of hope in the global fight against fascism. We treat Ethiopian veterans with great respect and take part in joint memorial events including annual celebration of the Patriots Victory Day. (Tkachenko, 2018)

Again, we see the word "respect" this time joined by "mutual trust." As well as Lavrov pointing out the "closeness of our views." This falls in line with the fact that all three countries selected for this study, as well as the majority of African countries, have conservative political positions, often at odds with the ideals supported by Western powers: such as the social inclusion of the LGBTQ+ community, gender equality, and democratic processes of governance.

These political alignments are also steeped in religious worldviews. Russia shares religious and social worldviews with Egypt, Ethiopia, and Zambia. Egypt consists of a mainly Muslim country, Ethiopia a majoirty Orthodox country, and Zambia is also majority Christian country. After Orthodoxy, Islam is the most popular religion in Russia, with close to twenty million followers (Husni et al., 2020). Several scholars have written about the long legacy of Islam in the region, and how the Kremlin is utilizing Islam in its geopolitical agenda (Hunter et al., 2004; Husni et al., 2020; Malashenko, 2000). Most famous is the delicate situation in Chechnya where the Russia shifted positions back and forth from broadly linking Islam to terrorism, to a more equanimous supportive role in favor of a politically advantageous

Pro-Russian (and also majority Muslim) Chechnya (Smirnov, 2006). Laqueur (2009) argues that "Muslim countries are natural allies against the West" (Laqueur, 2009), which at first glance appears a minimizing and divisive statement. However, one must consider the legacy of the USSR and how geography influences politics.

The USSR was made up of fifteen republics, six of which had a majority Muslim population, as well as a large Muslim population living in the Caucuses, as well as the Muslim Tatars historically located in Siberia. Although the early days of the Soviet Union disavowed religion (leading to violent purges), the government allowed limited religious activities in those six Muslim republics, understanding that they needed to play down atheist ideology in favor of political and geographic consolidation. Due to this maneuvering, at one point, Muslims enjoyed more religious freedom in the USSR than Orthodox Christians (though that was a short-lived experience). Eventually the USSR relaxed the position on religion, allowing more freedom in 1989, along with other more democratic reforms during the waning days of the empire (Bennigsen, 1985; Islam, 2014; Kemper, 2009).

Textual analysis of a newsletter from the archives of the Institute for African Studies within the Russian Academy of Sciences provided the following quote from Alexey Vasiliev, the Institute's director:

> The elections gave representatives of various sectors of the Egyptian society and various ideological convictions an opportunity to express their will and demonstrate their preferences. It turned out that most Egyptians reject both the secular liberal-democratic or left-wing parties that propagate Western values, adapted to some degree, to the specifics of the Egyptian society and its deep religiosity...The voters rejected the former ruling National Democratic Party and its ideology. The majority of Egyptians have tied their hopes for a better future, justice, political freedoms, higher living standards, the accountability of the government and return of dignity and national pride with Islamist parties. (p. 1, Inafran Archive Newsletter Volume 2, Issue 2, May 2012)

In this quote, Vasiliev is discussing the recent 2011 Arab Spring, which saw the ousting of former President Mubarak. As discussed earlier in this chapter, Mubarak was educated in the USSR, and in the 1980s, early in his presidency, expressed support for rekindling the Nasir era relationship between Egypt and the Soviet Union. However, he ended up a darling of the West, closely aligned with the United States for much of his 30-year presidency. So here, in this quote published not long after the deposing of Mubarak, Vasiliev supports this regime change, framing Mubarak as the secular West, with Egyptians campaigning for a religious East.

A critical discourse analysis of google reviews from the RCC in Egypt lends a complementary perspective, one student writing that the RCC in Cairo is "respectful and cultural" (review posted on June 22, 2020). Again, we see the word "respect," a word used in the Russian and Ethiopian data as well. However, earlier the same year, there are 30 reviews posted on the same day, seven of them negative, and three of those reviews posted each using the word "respect" negatively:

> *Review #1:* (translated from Arabic) "Customer service, Department of Languages, the utmost lack of professionalism or respect in dealing."
> *Review #2:* (translated from Russian) "these are not friends"
> *Review #3:* (translated from Arabic) "To be honest, a way of dealing is not respected by the security guards!! With visitors the reputation of the center greatly hinders!!"

All three reviews hint at cultural clashes within the RCC in Cairo. The third review is personally familiar, as I remember as a Black American, the conflicts I experienced with Russian security guards when I lived in St. Petersburg, the interactions reeked of hostility and xenophobia, much more than my experiences with other Russian people during my time there. What is interesting is that throughout the reviews from the RCC in Cairo, the positive reviews often had to do with the computer science and information technology courses, whereas the department of languages, received the majority of complaints.

One student wrote: "the service has a bad experience and treated like asphalt and the management of the Department of Languages are all rude." (review posted on November 4, 2019). Considering the fact that language and culture are meant to be the tools of soft power, the professionalism or lack thereof seems to have a detrimental effect. It could be racism, religious prejudice, or xenophobia, whichever the cause, it is leading Egyptian students to seek out the content knowledge of the sciences, rather than ascribing to a shared cultural connect of the Russian language. It is unclear from the reviews, whether Egyptian RCC students plan to transfer to Russian universities.

The concept of value homophily and transnational capital are not always solidified by religious camaraderie. These cultural clashes presented in the Egyptian data can also be found in the archival data about Ethiopia. Prior to the Soviet era, Emperor Menelik II harbored less romantic feelings about the Russians. Rupprecht (2018) states:

> He needed modern technology and knowledge to preserve Ethiopian independence, and he liked Russia for other pragmatic reasons: as a source of weapons, because Russia seemed not to partake in the European conquest of Africa, and because Russia was an absolute monarchy. When he decided to send young Ethiopians abroad to study, he therefore thought Russia more suitable than republican France. (Rupprecht, 2018, p. 219)

Here, Menelik engages in critical geopolitics, levying support for Russia vs. France, based on whether these respective countries shared opinions on governance and politics. The echoes of this sentiment reads in the legacy of Ethiopian and Russian relations, with Demassie (2021) stating during a recent Wilson Center panel: "Ethiopia's perception is that it can rely on Russia as an ally 'when Ethiopia doesn't get what it wants from the West'"(Demassie, 2021).

The combination of value homophily and transnational social capital encouraged and continue to catalyze the development of an elite class of Africans. However, in the case of the RCC in Egypt, a lack of dignified treatment and respect may hinder the socioemotional connections between African students and their Russian instructors. Ultimately, the impact on Egyptians studying abroad in Russia, cannot be assumed. African students from all three countries engage in critical geopolitics in their strategies both in studying abroad and in their home countries, ultimately choosing their best interests while simultaneously seeking to improve African human capital.

Research Methodology

This book was adapted from dissertation research approved by the UCLA Institutional Review Board (IRB) under the purview of the Office of Human Research Protection in 2020.

All names in this book are pseudonyms, assigned to protect the identity of the research participants. This work is the result of qualitative methodology including archival research, content analysis, a focus group, interviews, and observational methods. These various methodologies reveal the nuances of the ways in which policy actors construct meaning and act according to the constraints of institutional change (Hathaway, 1995).

Although the public health limitations of the COVID-19 pandemic required a heavy reliance on digital sources, I pursued a triangulation of data, using multiple methods of analysis and data sources to increase the credibility of findings (Denzin, 2012). A combination of qualitative and historical methodology allows for a complex understanding of the context of the problem, as well as shifting patterns of engagement over time, with a Pan-African focus on Russian educational initiatives and their impact on the continent. These methods guided the following research questions:

1. How was the Soviet Union involved in higher education development in Africa?
2. How is the Russian Federation involved in higher education development in Africa?
3. What is the purpose of Russian Cultural Centers (RCCs) in Africa?

The archival research provided the historical and political context of the problem. From the content analysis I gleaned the institutional goals, the intended educational outcomes as well as the student experience at RCCs. From the interviews, observation, and focus group I investigated the social phenomenology, and sociocultural contexts of the purpose of RCCs, as well as the student experience. As textual analysis is the main method used in critical geopolitics research, this combination of methods is an invaluable resource. This chapter will describe the data sources, collection procedures, and analysis process, closing with the limitations of the study and the positionality of the researcher.

Data Collection Procedures

Rather than investigating all nine RCCs located in eight countries in Africa, I chose three countries to examine: Egypt, Ethiopia, and Zambia. I chose these countries due to their geographic diversity: Egypt represents North Africa with two RCCs located in the cities of Alexandria and Cairo. Ethiopia represents East Africa and the Horn of Africa. Zambia represents both South and Central Africa in Africanist scholarly tradition. In addition, all the chosen countries have robust online data available, which was necessary while collecting data during the COVID-19 pandemic. One limitation of this research is the lack of representation in West Africa, however as of 2021, there are no RCCs located in West African countries.

I also chose these countries as they have varying cultural histories, specifically following Edmond Keller's theorizing on the differences between

RESEARCH CONSIDERATIONS & THEORETICAL APPLICATIONS

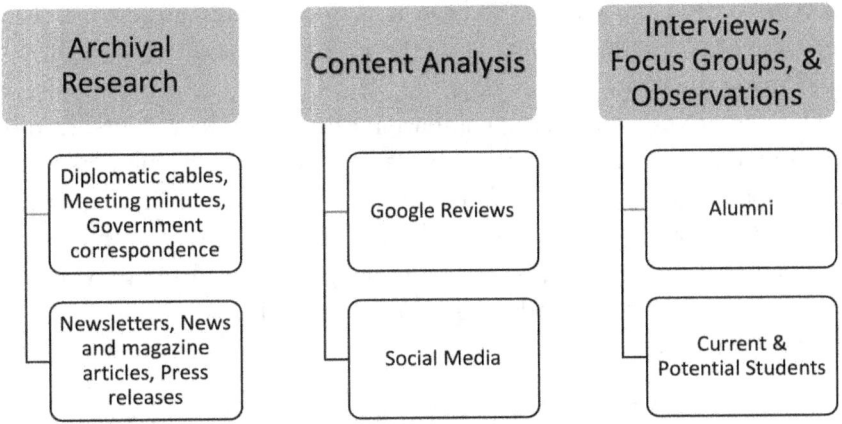

Figure 2. Data Sources

Indigenous Afrosocialism and Marxist-Leninist ideology. Ethiopia has a Marxist-Leninist influenced history, whereas Zambia is firmly indigenous afrosocialist, with Egypt providing an example of Arab socialism from the Pan-Arabism tradition (Keller, 1984; Pitcher & Askew, 2006; Torrey & Devlin, 1965).

The data for this research is organized into three categories: archival research, content analysis, and in-depth sociocultural data. The archival research included analysis of a variety of government, university, and press documents. The content analysis included google reviews and social media analysis, and the in-depth sociocultural data included interviews, a focus group, and observations of an event. In the following three sections I will describe each method and the associated data sources.

Archival Research

Documents are "social facts which are reproduced, shared, and used in socially organized ways." (Atkinson et al., 2001). Document analysis fields data as excerpts, quotations/passages that can then be organized into major themes, categories, and case examples through content analysis (Labuschagne, 2003). Bowen (2009) describes the five functions of documents when used alongside interviews: to provide participants' operational context, to suggest questions or situations to investigate, to track changes and development to the phenomena over time, to supplement other data, and to verify evidence from other sources (Bowen, 2009). Although documents are social facts in that they exist in time and space as indicators of human communication, they are often strategically manipulated, as is the case of political and social propaganda during the Cold War.

This research examines a combination of types of documents, to provide layers and various perspectives on the facts. I reviewed documents gathered from various digital archives related to historical and current Eurasian and African educational partnerships. Archival research is the examination of primary documents and is a necessary component of understanding the historical context of the problem. I conducted a targeted search of seven different digital archives including:

1. Worldwide Diplomatic Archives Index, U.S. Department of State
2. Digital Fieldwork Archives, Georgetown University
3. Open Access Archives on Soviet History, Harvard University
4. Archive Grid, WorldCat
5. Institute of African Studies, Russian Academy of Science, Moscow, Russia
6. The Wilson Center Digital Archive
7. UCLA Digital Archives

Using Boolean search strategies, I performed a systematic search using terms related to the USSR, Russia, Egypt, Ethiopia, and Zambia. I critically examined a total of 394 documents. From the Wilson Center Digital Archives, including diplomatic cables, meeting meetings, high-level government correspondence, telegrams, memoranda, speech transcripts, reports, and declassified scans of *CIA Intelligence Daily*. The Wilson archives contained 204 documents about Egypt, 92 documents about Ethiopia, and 13 documents about Zambia. I also did a close examination of the 37 newsletters published by the Institute from African Studies of the Russian Academy of Science. The final category of documents I examined were newspaper articles, magazines, and official press releases, including 26 articles from Ethiopia, 12 articles from Zambia, and 10 articles from non-African sources. Although all of the close to 400 documents informed the conceptual understanding of this work, please see the appendix for a list of the specific documents quoted in this book.

Content Analysis

This research offers online Google Reviews as a new venue of understanding in educational research. There are currently over 3,000 Google Reviews of the

RCCs in Africa, and I have chosen a sample of the 1,000 reviews written most recently prior to September 1, 2020. Rather than pursue a multiple case study, this is a critical discourse analysis with a Pan-African focus, highlighting the four countries listed below.

I also examined the official Facebook pages and websites for each RCC, critically examining the official posted content from the RCCs, posts from the student population and the general public, and Facebook Reviews about the RCCs. Most Facebook reviews were numerical, with a star ranking from one star to five stars. I discarded a numerical quantitative approach in favor of analyzing each review that contains more than one word of text. Of 1,000 Egypt google reviews sampled, 264 of the reviews contained text, 64 of the Zambia google reviews (plus an additional 20 Facebook reviews), and 2 of the 5 Ethiopia reviews.

This method is not new, as the field of semantic analysis and sentiment analysis for opinion mining is regularly used in research for business development to offer insight on products and services (Alamanda et al., 2019; Islam, 2014; Liu, 2012). However, semantic analysis is a complex quantitative approach requiring training in linguistics, attempting use generalizability to assess reviewers' opinions, rather than seeking a deep analysis of concepts. This project instead employed qualitative critical discourse analysis to organize, categorize, and theorize on the topic at hand. I applied a method of analysis delineated by Yung and Munksgaard (Young & Munksgaard, 2018).

Analyzing these online reviews provided context to the subsequent interviews, which may have more honest information due to their anonymity. In addition to Google reviews from various African RCCs, I analyzed a variety of other documents relating to RCCs, including decrees, newspaper articles, press releases, program proposals, institutional reports, public records, institutional reports, and legal documents (among others) found in digital archives.

As the data on Egyptian RCCs is based on anonymous, overseas Google users, my approach depends on textual analysis of their commentary in Google Reviews, rather than conversation built on rapport in a traditional qualitative interview.

Table 1. Google Reviews

Cairo, Egypt	3,110 google reviews
Lusaka, Zambia	185 google reviews
Addis Ababa, Ethiopia	5 google reviews

In-Depth Sociocultural Data

To understand the history and context of Russian-African partnerships in higher education, I conducted one on one, recorded interviews and one focus group via Zoom remote videoconferencing with ten participants born and raised in the countries of Egypt, Ethiopia, and Zambia. There was opportunity for deeper analysis by combining these portraits with document analysis, archival research, and interviews with key figures who may not necessarily be involved with RCCs on the ground, as well as African students who currently benefit from this programming. Individual, semi-structured interviews were conducted in-person and via videoconference with participants recruited via purposeful and snowball sampling from the researchers' academic and professional colleagues, as well as RCC students who have posted publicly on social media.

The categories of participants included: African alumni of higher education programs from the USSR and Russia, and current students attending Russian cultural centers in Africa. Seven of the ten interviews were conducted via Zoom were with current Zambian RCC students, who I also recruited to participate in the focus group. Two additional interviews were conducted with Ethiopian alumni of educational programming in the USSR, both were conducted in-person in Los Angeles, California. One interview was conducted with a participant from Egypt (via Zoom) with a potential RCC/Russian University student. Six of the ten participants were male, and four of the participants were female. Five of the participants were teenagers, three of the participants were in their twenties, and two of the participants were over the age of sixty.

In addition to interviews and a focus group, I observed the Zambian RCC's debate team competition. It was two hours in length, livestreamed, videorecorded, and posted on the official RCC Facebook page. The interviews, focus group, and observation were all transcribed for a small fee by an online transcription service (rev.com).

Data Analysis

I analyzed a sample of the most recent 1,000 of the google reviews to pilot my approach. I maintained two spreadsheets, one for reviews that have actual text to analyze, and one spreadsheet for reviews that only have a number ranking. I also wrote analytic memos, and engaged in two levels of inductive coding, in preparation for data analysis. After my initial dissertation defense, I submitted

Table 2. Participant Demographics

Pseudonym	Affiliation	Age	Gender
Dave	RCC student, Zambia	21	Male
Violet	RCC student, Zambia	20	Female
Lena	RCC student, Zambia	18	Female
Mundo	RCC student, Zambia	19	Male
Tom	RCC student, Zambia	18	Male
Lucy	RCC student, Zambia	19	Female
Akaso	RCC student, Zambia	19	Male
Yonas	Alumni of USSR program, Ethiopia	61	Male
Abraham	Alumni of USSR program, Ethiopia	64	Male
Alana	Prospective RCC student, Egypt	25	Female

my proposal to my university's International Review Board and was approved to conduct interviews and host focus groups. I uploaded my spreadsheets into Dedoose qualitative analysis software for a more efficient and multifaceted analysis of the data. Dedoose was used for the Google reviews, interview transcripts and quotes from archival documents.

I followed the inductive approach of Yung and Munksgaard, who articulate a step by step plan for using qualitative analysis software (Young & Munksgaard, 2018). I have chosen inductive rather than deductive coding, as this approach is recommended in comparative study and case studies, to examine patterns that emerge across multiple cases (Fereday & Muir-Cochrane, 2006). Although this work is not a formal case study, it does examine RCCs in three different countries, each with their own historical context and institutional culture. Although the computer analysis software provides an inductive approach, the process was abductive—with continuous movement between categorizing and interpreting, a comparative approach, changing classifications as necessary, as I comb through the data (Timmermans & Tavory, 2012).

In framing the data analysis process there were three areas of examination: by Country, by Data Source, and by Stakeholder. Examining the data by country allowed me to understand the geopolitics that vary by region, the diverse cultural aspects impacting this inquiry, and the historical context of Russian relations that differ by country. Examining the problem by data source, allowed me to gain two different perspectives, the first: how Africans make meaning Russian educational opportunities, and the second: how the Russian government presents themselves to Africa.

The interviews, focus group, observations, and content analysis provided the first perspective, and the archival research and content analysis provided the second perspective. The third area of examination: Stakeholder, allowed me to delve into the different experiences that of African alumni of Soviet/Russian programs, current RCC students, the African media, and the Russian government. Each stakeholder holding a different view on the role of RCCs in Africa. This a formal case study or comparative case study, rather an examination of RCCs in three different countries, each with their own historical context and institutional culture.

Limitations

The limitations to this study include the fact that most of the data was collected online versus in person, due to the data collection period coinciding with the COVID-19 pandemic.

In addition there were varying amounts of data per region: for example more interviews from Zambia than from Egypt and Ethiopia. However, this disparity was compensated by the extensive archival data available on Ethiopia and the extensive google reviews on the RCC in Egypt. Other limitations included that there were no teacher/instructor perspectives in the data, and only one of the two RCCs in Egypt (Cairo, excluding the RCC in Alexandria) was studied.

Positionality

> My first taste of the feeling of "tribalism" happened when I came to the USA. I read books written by some anthropologists and half-baked western scholars about exaggerated differences among Ethiopian ethnic groups. I remember well one summer in the early 60s when a sociologist I met for the first time at an Ethiopian Peace Corps volunteers training camp, asked me "where do you come from?" I said "of course, Ethiopia" He said, "I know, are you Amhara, Oromo, Tigre?" I responded "Ethiopian." He then asked what language I spoke, I said, "I speak Amharic, Oromiffa, and some Tigrigna, and so on." "Ok", he said, "what is your religion, are you Christian or Muslim?" I responded, "I believe in the One Almighty God". The arrogant sociologist was frustrated with me and walked away. (Isaac, 2016)

In this quote, famous Ethiopian scholar and professor Ephraim Isaac circumvents incessant categorization, and in this way my positionality remains a simple one. This research was developed as a foreign observer of the critical geopolitics at play in Africa and Eurasia. The interest in these regions formed after living in St. Petersburg, Russia, first as a Fulbright scholar, and later as a professional musician. As a Black American, with Pan-African sensibilities and a tendency towards diasporic unity, my interest in Africa is both past—through ancestral connection, and future—in the projection of Africa as the future of economic, scientific, and technical innovation.

References

Al-Awadi, H. (2005). Mubarak and the Islamists: Why Did the "Honeymoon" End? *The Middle East Journal*, 59(1), 1–19.

Alamanda, D. T., Ramdhani, A., Kania, I., & Hadi, E. (2019). *Sentiment Analysis Using Text Mining of Indonesia Tourism Reviews via Social Media*. GI Social Sciences Forum.

Arafat, A. A.-D. (2011). *Hosni Mubarak and the Future of Democracy in Egypt*. Springer.

Åslund, A., Guriev, S., & Kuchins, A. (Eds.). (2010). *Russia after the global economic crisis*. Columbia University Press.

Atkinson, P., Coffey, A., Delamont, S., Lofland, J., & Lofland, L. (2001). *Handbook of Ethnography*. Sage Publications.

Bennigsen, A. (1985). Islam in the Soviet Union. *Journal of South Asian and Middle Eastern Studies*, 8(4), 115.

Berman, B. J. (1974). Clientelism and Neocolonialism: Center-Periphery Relations and Political Development in African States. *Studies in Comparative International Development*, 9(2), 3–25.

Bowen, G. A. (2009). Document Analysis as a Qualitative Research Method. *Qualitative Research Journal*, 9(2), 27.

Contributor. (2018). We Plan to Create an Ethiopian Center for Nuclear Science and Technologies. *The Reporter*. https://www.thereporterethiopia.com/article/we-plan-create-ethiopian-center-nuclear-science-and-technologies

Council, N. S., Sens, A. D., Kohl, H., Lake, A., Christopher, W., Holbrooke, R. C., . . . Clinton, W. J. (1995). *Memorandum for Kenneth C. Brill from Andrew D. Sens, 'Memorandum of Conversation of the President's Expanded Meeting with Chancellor Kohl of Germany'*. https://digitalarchive.wilsoncenter.org/document/209782

Demassie, A. A. (2021). *Global Perspectives: Ethiopia-Russia Relations* [Interview]. The Kennan Institute. https://www.wilsoncenter.org/event/global-perspectives-ethiopia-russia-relations

Denzin, N. K. (2012). Triangulation 2.0. *Journal of Mixed Methods Research*, 6(2), 80–88.

Fereday, J., & Muir-Cochrane, E. (2006). Demonstrating Rigor Using Thematic Analysis: A Hybrid Approach of Inductive and Deductive Coding and Theme Development. *International Journal of Qualitative Methods*, 5(1), 80–92.

Filatova, I. (2001). Russia and Africa. *Journal of African History*, 42(2), 347–348.

Gleberzon, W. (1978). Marxist Conceptions of the Intellectuals. *Historical Reflections / Réflexions Historiques*, 5(1), 81–97.

Gleijeses, P. (2002). *Conflicting Missions: Havana, Washington, and Africa, 1959–1976*. Univ of North Carolina Press.

Gyekye, K. (2015). Political Corruption: A Philosophical Inquiry into a Moral Problem. *Philosophy and Politics: Discourse on Values, Politics, and Power in Africa*, 10(2), 353.

Hathaway, R. S. (1995). Assumptions Underlying Quantitative and Qualitative Research: Implications for Institutional Research. *Research in Higher Education*, 36(5), 535–562.

Hunter, S., Thomas, J. L., & Melikishvili, A. (2004). *Islam in Russia: The Politics of Identity and Security*. ME Sharpe.

Husni, H., Akhmedov, O., Herlina, N. H., & Kormiltsev, I. (2020). Islam in Russia: History, Challenges, and Future Perspective. *Religious Studies: An International Journal*, 8(1), 45–66.

Hyslop, J. (2005). Political corruption: Before and after apartheid. *Journal of Southern African Studies*, 31(4), 773–789.

Jenkins, P., Tahaafe-Williams, K., Welby, J., Robert, D. L., Maxwell, D., Freston, P., . . . Kings, G. (2018). *Encyclopedia of Christianity in the Global South* (Vol. 2). Rowman & Littlefield.

Kebede, M. (2003). From Marxism-Leninism to Ethnicity: The Sideslips of Ethiopian Elitism. *Northeast African Studies*, 10(2), 165–188.

Keller, E. J. (1984). The Ethiopian Revolution: How Socialist Is It? *Journal of African Studies*, 11(2), 52.

Kotek, J. (2015). *Students and the Cold War*. Springer.

Kuraev, A. (2014). *Internationalization of Higher Education in Russia: Collapse or Perpetuation of the Soviet System? A Historical and Conceptual Study* [Dissertation]. Boston College.

Krupskaya, N. K. (1940). *On Education: Selected Articles and Speeches*. Gvardiya Publishing House.

Lange, M. K. (2004). British colonial legacies and political development. *World development*, 32(6), 905–922.

Labuschagne, A. (2003). Qualitative Research: Airy Fairy or Fundamental. *The Qualitative Report*, 8(1), 100–103.

Laqueur, W. (2009). Russia's Muslim Strategy. *Middle East Papers*, 1.

Law, I. (2016). *Red Racisms: Racism in Communist and Post-Communist Contexts*. Springer.

Lenin, V. I. (1920). Report of the All-Russia Central Executive Committee and the Council of People's Commissars on the Home and Foreign Policy to the Eighth All-Russia Congress of Soviets. Part II. *Report On The Work Of The Council Of People's Commissars, December*, 22 .

Lenin, V. I. (2012). *Essential Works of Lenin:" What is to be done?" and Other Writings*. Courier Corporation.

Leviyeva, E. (2005). The Changing Face of the Russian Democracy: Racism and Xenophobia in Russia-Foreign Students under Attack in Russia and US. *Rutgers Race & L. Rev.*, 7, 229.

Lilge, F. (1968). Lenin and the Politics of Education. *Slavic Review*, 27(2), 230–257.

Lubinda, N. (2015). *Given Lubinda's Daughter Says it's Not Her Father's Problem That Students in Russia Are Starving*. https://www.zambiawatchdog.com/given-lubindas-daughter-says-its-not-her-fathers-problem-that-students-in-russia-are-starving/

Makasa, E. (2008). The Human Resource Crisis in the Zambian Health Sector–A Discussion Paper. *Medical Journal of Zambia, 35*(3), 81–87.

Marchetti, V., Marks, J. D., & Wulf, M. L. (1974). *The CIA and the Cult of Intelligence.* Knopf New York.

Marx, K., & Engels, F. (1964). *Karl Marx and Friedrich Engels on Religion.* Schocken Books.

Nash, M., Cowcher, K., De Oliveira, A. B., Ribeiro Sanches, M., Savage, P., Siegert, N., Vasić-Janeković, V. (2016). *Red Africa: Affective Communities and Cold War.* Black Dog Publishers.

Ojo, J. S. (2018). Politics of corruption in Africa. *Encyclopedia of Public Administration, Public Policy, and Governance.* New York: Springer International Publishing. https://doi.org/10.1007/978-3-319-31816-5_3556-1

Pierce, S. (2006). Looking like a state: colonialism and the discourse of corruption in Northern Nigeria. *Comparative studies in society and history, 48*(4), 887–914.

Po-Keung, H., & Tak-Wing, N. (1999). Comprador Politics and Middleman Capitalism. In Ngo Tak-wing (ed.), *Hong Kong's History* (pp. 30–45). London and New York: Routledge.

Profile of Nkandu Luo. (2021). National Assembly of Zambia. Retrieved June 27, 2021, from https://web.archive.org/web/20171014183901/http://www.mgcd.gov.zm/index.php/about-us/ministers-office

Rupprecht, T. (2018). Orthodox Internationalism: State and Church in Modern Russia and Ethiopia. *Comparative Studies in Society and History, 60*(1), 212–235.

Russell, M., & Pichon, E. (2019). *Russia in Africa: A new arena for geopolitical competition.* EPRS Briefings, November.

Schmidt, E. (2013). *Foreign Intervention in Africa: From the Cold War to the War on Terror.* Cambridge University Press.

Sklar, R. L. (1979). The Nature of Class Domination in Africa. *The Journal of Modern African Studies, 17*(4), 531–552.

Small, N. (1977). Zambia–Trouble on Campus. *Index on Censorship, 6*(6), 8–14.

Smirnov, A. (2006). The Kremlin's New Strategy to Build a Pro-Russian Islamic Chechnya. *Eurasia Daily Monitor, 7*(9).

Timmermans, S., & Tavory, I. (2012). Theory Construction in Qualitative Research: From Grounded Theory to Abductive Analysis. *Sociological Theory, 30*(3), 167–186.

Tkachenko, V. (2018, April 7). *Ethiopia and Russia: Kindred Spirits* [Interview]. The Reporter. https://www.thereporterethiopia.com/article/ethiopia-and-russia-kindred- spirits

Tödt, D. (2012). *Les noirs perfectionnés.* Humboldt-Universität zu Berlin, Philosophische Fakultät I.

Weaver, H. D., JR. (1985). *Soviet Training And Research Programs For Africa* (Order No. 8509613). Available from ProQuest Dissertations & Theses Global. (303414728). https://www.proquest.com/dissertations-theses/soviet-training-research-programs-africa/docview/303414728/se-2

Young, L., & Munksgaard, K. B. (2018). Analysis of Qualitative Data: Using Automated Semantic Analysis to Understand Networks of Concepts. In *Collaborative Research Design* (pp. 251–284). Springer Singapore. https://doi.org/10.1007/978-981-10-5008-4_11

· 4 ·

THE NEW RED SCARE

Abstract: This chapter traces the slew of recent military and land use deals between Russia and Africa, connecting these treaties with recent educational initiatives on the continent. This chapter also offers vignettes from African students, who describe a waning interest in the United States and the United Kingdom, and an increasing interest in studying or working in Russia, China, and India. Expanding beyond Nye's soft power, this closing chapter considers how religion, sexual orientation, gender expression, and political ideology challenge notions of North and South, shaping and developing into an increasingly multipolar world. This chapter considers the present and the future of engagement between Africa and Eurasia, highlighting the new economic and military partnerships between Eastern Europe and Central Asia, the encroaching power of China, and how Africa and the African diaspora play a critical role at the center of this discussion, wielding increasing power on the geopolitical stage.

Keywords: Geopolitics, Diplomacy, Intelligence, Relations, Education, Higher Education, International Education, African Education

Introduction

During the course of his long life W.E.B. DuBois travelled to many countries throughout Europe, Asia, and Africa. Two years after the trip to the USSR where he met with Khrushchev—and was influential in brokering Pan-African and Russian cultural relations—he also visited China. DuBois had previously visited the China in 1936 and returned in the late 1950s spending eight weeks travelling throughout the country.

Young and Green (1972) note that DuBois visited not long after then President Nixon had travelled there, and he was treated as a foreign dignitary at the same level of luxury and respect as Nixon. Yet, the American media published no mention of DuBois' trip and his positive experiences in communist China. It is of note that a Black academic and political activist was treated as id he was as important to China as the sitting U.S. president. (Young & Green, 1972). The FBI most certainly considered his visit a threat, when reviewing the now declassified FBI file on DuBois. In the 66-page FBI file (the last part of a series of five files on his life), there are several notes on this trip to China, quoting DuBois' praise of the will, political consciousness, and industriousness of Communist China (FBI, 1959).

In this chapter I offer a discussion of this new phenomena of Russian Cultural Centers as the culmination of centuries of both Soviet and Russian higher education development. I also offer conclusions and implications for American, Russian, and African audiences, as well as promising areas for future research. I close this work by looking to the future, and the role China may play in engaging higher education as critical geopolitics. One might consider Russia to be the past-present, and China to be the present-future in terms of geopolitical partners engaging with the African continent.

The Purpose of Russian Cultural Centers in Africa

State Perspectives

Much has been written about the "end of ideology" upon the Russian government's transition to a capitalist oligarchy (Knight, 2006; Light, 2003; Lipset, 2017; Morris, 1994). In this way, the purpose the current Russian higher education agenda is devoid of the seeds of Marx, Engels, and Lenin. According to the perspective of the State, the purpose of RCCs in Africa is to gain access to (from Russia's perspective) or gain income from (from African governments' perspective) valuable natural resources. The second purpose of RCCs in Africa is to cultivate political support in both directions: Russia seeking political support in Africa, and African nations seeking a partner in Russia.

Although the USSR expressed a vision of supporting newly liberated African nations in defiance of their colonial masters, their soft power higher education development hints at neocolonial intentions of their own. For example, natural resource seeking in exchange for training began in the 1930s with extensive geological exploration from specialists who travelled to Angola, Benin, Ethiopia, and Mali, building laboratories and national geological centers and training local technicians. This later led to the construction of power plants and hydroelectric stations in the 1960s, such as the Malka Wakana in Ethiopia, and the Aswan Dam built across the Nile river in Egypt. (Kochetkova, 2009).

Today, Russia trades and invests significantly in the energy and mining sectors in Africa, specifically: oil, natural gas, diamonds, gold, aluminum, nickel, and copper in twenty African countries all with recent trade agreements signed in June of 2018 (Russell & Pichon, 2019). On the topic of mining and energy field operations, Hedenskog (2018) highlights some geographic barriers and bilateral solutions:

> Such partnerships bring African countries the capital and know-how they need to tap into their energy potential. For their part, Russian energy companies gain an opportunity to expand production at lower cost than in Russia, where many of the country's untapped reserves are difficult to access, being located under deep water or in Arctic regions.... Despite its own huge mineral resources, Russia has some critical shortage of certain raw materials, including chrome, manganese, mercury, and titanium, and faces depletion of reserves of others, including copper, nickel, tin, and zinc. It also needs coltan and rare earth metals for new technologies. (Hedenskog, 2018)

In a 2008 newsletter from the archives of the Institute for African Studies, Alexey Vasiliev, the director of the Institute of African Studies in Moscow, does not mince words or sugar coat the purpose of Russian engagement in Africa, stating on page 14 of a 2008 newsletter:

> The visits of President V. Putin to a number of African countries—Egypt, the Republic of South Africa, Morocco, Algeria, and Libya—have demonstrated Russia's growing interest in Africa. Our country needs some raw materials mined in Africa and its agricultural products and wishes to develop trade relations with the countries of the African continent.

Related to accessing natural resources, is Russia's ongoing attempts to build military bases on the African continent. During the Cold War, Russia was granted access to naval bases in Algeria, Egypt, Ethiopia, Guinea, Libya, Somalia, and Tunisia. Due to their support of Angolan liberation, Russia maintained military advisors and selected troops at the military base in Luanda for 25 years. In this quest for land access for military endeavors, Russia seems to be operating even in possible conflict with their traditional partner: Ethiopia, cultivating partnerships that encroach on the boiling tensions in the Horn of Africa:

> If Russia is looking to strengthen its ability to sustain naval deployments in the Red Sea, Gulf of Aden, and western Indian Ocean, [an alternative to Sudan] could be Eritrea. In September 2018...Eritrean foreign minister Osman Saleh met with Russian foreign minister Sergey Lavrov in Sochi. The parties signed an agreement which suggests an emerging commercial-military relationship including the establishment of a logistics center in the Eritrean port of Assab.... Russia has also had contacts with the breakaway region of Somaliland. In exchange for establishing a small multiuse air and naval facility in the Djibouti-bordering town of Zeila, Russia would formally recognize the region's 'independence' from Somalia. (Hedenskog, 2018)

A report published by the Swedish Defense Agency in 2018 states that twenty-one African countries have recent military cooperation agreements with the Russian Federation. These new agreements signed between the years 2015 and 2018, during the time of western sanctions against Russia; on that list are Egypt, Ethiopia, and Zambia. For each country, the report lists the main purpose of the agreements, which varies according to each country's needs.

For Egypt, an agreement was signed in November 2017 "deliveries of equipment and weapons for counter-terrorism operations." For Ethiopia, an agreement was signed in April 2018, for "training and cooperation on peace-keeping and counter-terrorism and antipiracy efforts. Deal not yet in force." For Zambia, and agreement was signed in April 2017 for "provisions for the supply of weapons and delivery of spare parts" (Hedenskog, 2018). However, these military agreements are not supported by all citizens of these countries. Tribal leader Edith Nawakwi expressed strong opposition to the sales and trade agreements made between Zambia and Russia, specifically opposing the sale of Russian MI17 military helicopters to the Zambia government. Nawakwi highlights the obvious fallacies present in the isomorphic behaviors of the State, when she challenged President Lungu to explain the purchase:

Those aircrafts are used in places [where] you need to carry 32 armed military personnel who can be dropped here and there. Or in fact those are aircrafts for countries which have money because to run a MI17 you need $3,000 per hour and what has baffled me is that this government has gone to Russia and contracted those useless helicopters, five of them for Zambia Police and one wonders where Zambia Police will park them [and] where they will get the pilots since ZAF doesn't even have pilots and the technicians who are qualified enough to run them. (*Nawakwi questions*, 2015)

Nawakwi then gets the heart of the controversy when she identifies the critical geopolitics at play, stating:

They are bringing them here purportedly for the police under the Russian-Zambian Government debt swap. When I was Minister of Finance, we would use the debt swap to provide for social services for children like the ones we have just met. But because these people are bent on pocketing resources, they go to Russia and collect those aircrafts.

In addition to accessing natural resources and land for military occupation, Russia seeks to cultivate political partnerships with likeminded conservative countries. In a 2018 interview published in an Ethiopian newspaper, Ambassador Vsevolod Tkachenko said:

The success of Russian-Ethiopian political interaction is based on identical or similar positions on major global issues. Our Ethiopian partners consistently support Russian initiatives in the United Nations on the basis of reciprocity. (Tkachenko, 2018)

"Identical" is a strong word to use when comparing the world views of an African country and a Eurasian country. But this serves the purpose of this interview, in which Tkachenko applies a thick layer of propaganda. Notice, here the use again of the word "respect" two times in his statement, as well as mentioning culture and tradition:

Today we are witnessing a desire of certain Western countries to create a unilateral world model according to their own scenarios, disregarding cultural features, traditions, [and] development level of other nations and peoples. In doing so, they often flagrantly violate fundamental principles and norms of international law, [and] show lack of respect for world community. From the Russian point of view, this approach is unacceptable since it destroys traditional values developed in societies for generations. In countering these dangerous trends, we feel support of many countries including Ethiopia. We all agree that the only possibility of further development is to create and strengthen a genuinely multipolar world based on principles of mutual respect and balance of interests.

Tkachenko highlights (in so many words) the concept of value homophily incentivizing the continued allegiance of Ethiopia in geopolitical affairs (Tkachenko, 2018). Not only has Russia expressed support for controversial partners in Africa, there has also been a reciprocal effort to protect each other's interests. During the 2014 UN General Assembly Resolution opposing Russia's annexation of Crimea, 25 out of 54 countries in Africa abstained from voting, and two countries opposed the resolution. Later in 2016, Russia withdrew from the Rome Statute which planned to establish an International Criminal Court, citing their support for the Africa countries also withdrawing from the statute (Russell & Pichon, 2019).

> For many African countries, Russia's willingness to ignore human rights problems and offer no strings attached political and military support makes it a welcome ally... Russia has used its UN Security Council veto to shield African countries, such as Zimbabwe in 2008, from international human rights-related criticisms. (Russell & Pichon, 2019)

Albert Kofi Owusu, who is the General Manager of the Ghana News Agency, a state-run media source, articulates the impact of value homophily, stating:

> With Western aid, there are all these conditions. They say: 'if you want this money, you have to do this about LGBTQ', for example—even if it goes against your country's values. China and Russia say, 'here's the money,' and that's it. (Kulenova, 2019)

In conclusion, the shared conservative political and religious beliefs between Africa and Eurasia trump the historical legacy of socialist ideology.

Student Perspectives

For African alumni of Russian and Soviet universities, RCCs are place for social engagement and nostalgia. For currently enrolled and potential future RCC students, the purpose of RCCs in Africa is to provide language and cultural training in order to attend college abroad in Russian universities. This training cultivates transnational social capital bolstered by value homophily, in order to develop and improve African human capital.

Interviewing Zambian RCC students, when asked why they decided to attend a RCC, some stated it was because of an advertisement they saw online, others because of a childhood fascination with the Russia portrayed in films. However, most of the students enrolled in the RCC in Lusaka because they either had a family member or a friend who previously studied abroad in the

Soviet Union, or currently study abroad in Russia. During the focus group, Akaso, a nineteen-year-old RCC student from Lusaka stated:

> I've got my cousin who is in Russia. He's the one who tells me about Russian culture, and he told me you know, if you'd like to get more opportunities, I advise you to go to the Russian culture center. You apply [for college/a scholarship] and then one day you become someone. (Akaso, Zambian RCC Student Interview, April 30, 2021)

As discussed earlier in this chapter, 80 % of RCC students who apply to scholarships to Russia, receive them. This indicates the power of transnational social capital—not only did family and friends catalyze these students' interest, they also provided the insider knowledge that led to tangible, fiscal rewards. For Dave, a twenty-year-old Zambian student, the RCC provides an education, professional development, and a social circle. Dave spends 32 hours a week at the center in Lusaka, attending Russian language class, in addition to participating in various center-hosted clubs, including the Culture Club, Chess, Debate, Poetry, Music, Public Diplomacy, and Sports Club (which includes soccer, tennis, and badminton).

Dave's active engagement in the club led to a supporting role managing the official social media pages for the Lusaka RCC. With an interest in studying information technology followed by a career in entrepreneurship, this role provides him with professional development. Multiple Zambian RCC students expressed a desire to pursue a future career as an entrepreneur, planning to study in the fields of Economics, Agronomy, Nuclear Physics, and Mining Engineering. Tom, an eighteen-year-old RCC student from Lusaka was one of several who broached the subject of brain drain, brain gain, and brain circulation, although he did not use those terms. Tom states:

> In Zambia we do not have so much knowledge about nuclear physics because obviously Zambia is a third world country. Another reason why I decided to go to Russia is because here in Zambia, we're not offered the opportunity to study nuclear physics and technologies, but in Russia, it's there. So obviously you gain some knowledge outside and come back here and implement it. (Tom, Zambia RCC Student Interview, April 30, 2021)

Lucy echoes this sentiment, stating:

> It's just for our government to put in their A-game and try to partner up with other people, not only one, there's more ideas. So I think they should also put in more incentive for the children so they can learn more, so that as they come back to the

country, they can also be able to teach the others as well. (Lucy, Zambia RCC Student Interview, April 30, 2021)

Violet, near the end of the focus group, also touches on this topic, saying:

> I would like to contribute something. In the hope that the government is sending their youth to go study abroad, it is up to the youth society to keep it in mind while studying abroad that we have to come back to our countries because sometimes most of us would want to stay back. Maybe the life is just so amazing, maybe you want to stay back, just don't want to go back home. I think the huge role that's big that shouldn't be played with is by us, the youth, trying to remember that we are there for our country, and afterwards we have to come back and not just keep the ideas on that side, but to bring it here. (Violet, Zambia RCC Student Interview, April 30, 2021).

One of the interview questions that arose after learning of RCC students' familial connections to study abroad in Russia, probed at what expectations the students had for when they eventually studied in Russian universities. Lucy, and eighteen-year-old, articulates the vital necessity of a transnational network, stating:

> I think it's going to be a very good experience. And then for me, what I was told is that Russians are busy people, being people who mind their own business. So for us here in Zambia it's when you're moving in the road, you can easily say hi to when you meet. But my friends told me to say, in Russia they are busy they don't really have that time to stop someone and want to say hi to them. So I feel such information is very helpful to us so we know what we are not supposed to do when we get there, and what are supposed to do when we get there. So having people there is actually a good thing for us here, so for us I think when we go there, we will also help the people coming behind us to say this is what you're supposed to do. So it's good for us.
> (Lucy, Zambia RCC Student Interview, April 30, 2021)

Lucy highlights both the benefits she receives from her network abroad, and how she plans to return the favor, to the next generation of students. Lena agrees with Lucy, stating:

> My sister is there, and I like the things that people say about Russia, they're such cute people. She say it's quite peaceful and people are quite welcoming, you just have to mind your own business. And you don't really have to be so sensitive because the people haven't seen Black people...when they see you, they'll be like, oh wow a Black person! [everyone laughs] That information is really important.
> (Lena, Zambia RCC Student Interview, April 30, 2021)

Akaso jumps in with why he believes studying abroad will be a good learning experience, unknowingly referencing the benefits of cultural competency, stating:

> Going abroad makes one know a lot of things, not just to be limited in one thing, not only that's you, your tribe then different cultures. For example, Russia is not the way you think they really portray themselves, maybe in the industry, whatever they may be. So you try to learn different things, maybe their culture, the way they dress, the way they attempt work...So we tend to learn a lot of things. (Akaso, Zambia RCC Student Interview, April 30, 2021)

In a focus group, seven Zambian students were asked why they believed Russia built an RCC in their country, and why their government allowed it. Dave answered:

> I think we do have Russian cultural centers in our country because Russia and Zambia go way back. When Zambia gained its independence in 1964, Russia was actually the first country to congratulate Zambia...I think through that, they showed they had great interest in our country, great respect and they really wanted us to be friends and think our bilateral relationship has just always been. (Dave, Zambia RCC Student Interview, April 30, 2021)

Again, note the use of the word "respect" a pervasive theme throughout this research inquiry. Lena also states her appreciation for Russia:

> And I just want to say education-wise: Russia has been helping us next day. It's one of the countries that had given us the most information, and technology wise. So I like Russia, if a student goes to Russia to gain more knowledge and more experience. And then come back to that Zambia. I think it would be way better. (Lena, Zambia RCC Student Interview, April 30, 2021)

Dave, like his peers was not ignorant to the political implications of transnational education agendas, discussing the competition between China and Russia:

> I do believe Africa [and] part of Asia is going to be the next future because I feel like at this point these countries are beginning to follow countries. Like well, some from the Western side, some from Europe, just because of the tension that is happening amongst each other... I think the future economies is definitely China because those guys are already way ahead of everyone else. But I think agriculture wise definitely Africa.

When pressed to discuss the impact of China on Africa Dave responded:

> Basically they are helping us build institutions, basically our infrastructure they're helping us with it, but I think as Einstein say, "for every action is an equal and opposite reaction." So I guess there's like they're helping us with some stuff, but then there's always negatives to it. (Dave, Zambia RCC Student Interview, April 30, 2021)

And after the follow up question of would you accept a job from a Chinese employer or a Russian employer, Dave answered that he would prefer a Russian employer due to his knowledge of both the language and the culture. Lena also brings up the topic of China:

> Currently we have our economic crisis. Our economy is quite really low. Compare the dollar and our quarter its really bad. So I feel like Russia it could play a role, yeah, for now there is more coordination with Chinese people. And if you come to our country, we have more China things than Russian things. So I feel like I think the one thing that would help us as culture. If we grow more, maybe we increase our economy. (Lena, Zambia RCC Student Interview, April 30, 2021)

When asked about attending other cultural centers in their country, such as the Italian and French cultural centers, the students expressed little interest, with a laser sharp focus on the educational and employment trajectory promised in enrollment in the RCC. Similarly, these types of institutions, even though housed in foreign countries, often employ the same racist and xenophobic tactics used in their home countries. Dr. Elizabeth Giorgis, an associate professor at Addis Ababa University, describes the disrespect she experienced while curating and exhibit at the German Cultural Center in Ethiopia:

> There was, for instance, an episode that really perplexed me. I was accused of manipulating the museum space I directed...A complaint was filed by an artist and curator to the German Embassy and the German cultural institute instead of apprehending my Ethiopian superiors. I was flabbergasted!! Since the museum space was close to the German cultural center, both artist and curator perhaps thought I should be reprimanded by the proximity of whiteness. How can the artist and curator deny the conditions of hierarchy, racial or otherwise? And it was with great pain, a pain that I still deeply feel, that I wrestled this incoherently silenced episode of my return. I complained about this event but no one seemed to understand my moan. (Giorgis, 2020)

Upon online perusal, the German Cultural Center in Ethiopia, called the Addis Ababa Goethe-Institute, has a much larger budget and online presence than the RCC in that city. However, it looks to be affiliated with a local university, the Sedist Kilo Compound of College of Business and Economics, whereas the RCC operates independent of any accredited local institutions. Germany, a prominent representative of Western power, represents here, a declining African interest in the West.

In the interviews with current Zambian RCC students, when asked if they would consider studying abroad or working abroad in the United States, Canada, or the United Kingdom, none expressed any interest. Only one of the seven students was even interested in vacationing in the U.S., followed by two who were interested in vacationing in the UK or Canada. Analyzing the archival data on Ethiopia, even the connotation of the words chosen by Ethiopian journalists reflects a shift in perception of the U.S. versus Russia. For example:

> In the shadow of the official visit by the United States Secretary of State, Rex Tillerson, to the nation, the Russian Foreign Minister, Sergey Lavrov, made a working visit highlighting a new partnership with Ethiopia. On areas of trade, nuclear development and academic scholarships for Ethiopians, the minister had a busy two-days in meetings with the leadership of Ethiopia. (Getachew, 2018)

From the very first words in the sentence "in the shadow of" we see the way Ethiopia views the United States, or at least, the way the media and general public might view the U.S. Indeed, in interviewing Ethiopian alumni, Zambian RCC students, and an Egyptian potential student, representatives from all three countries noted the current instability of the U.S. and the waning days of Empire. Surely the media representations of the U.S. are a concern for foreign students, seeing and reading about the combination of civil unrest, state-sanctioned racism, and economic hardship in America.

There are several key takeaways from this study, the first is that soft power-fueled educational initiatives must function as more than just centers of spreading hegemonic culture.

In the case of all three RCCs in Egypt, Ethiopia, and Zambia—African students are seeking the best possible strategy to reach their career goals. The Egyptian RCC students were often unhappy with the interpersonal experiences with Russian staff in the Languages department. But, the value of the information technology/computer science courses offered there, were an incentive to continue enrollment.

For Zambian students, studying the Russian language means access to the best nuclear physics programs in order to bring that knowledge back to Africa—rather than language study as a first step towards permanently immigrating to Russia. The network developed at the Zambian RCC, was invaluable in the scholarship application process, with the RCC providing transnational social capital and insider knowledge on the bureaucracies of higher education admissions for international students.

Ethiopian RCCs serve more as a physical touchpoint of cultural nostalgia, with many Ethiopians suspicious of the political agenda of the Russian Federation. Ultimately, in each country, the RCC served various purposes despite the original mission and intent of the organizations. Meeting the population at their most vital need: whether it be IT training, social engagement, or the development of a transnational network, should the goal of policy and practice in informal higher education.

Areas for Further Study

This book provides an overview of the historical and political context and impact of RCCs in Africa. However this is not an exhaustive inquiry, as there is always room for more study. Future research should include comparative case studies of all nine RCCs in Africa, rather than focusing on the three countries of Egypt, Ethiopia, and Zambia. Another research project could include a comparative case study of Egyptian Russian University and Egyptian Chinese University, examining the competing motivations of major political powers currently engaged in African development.

In addition, there is the opportunity to pursue longitudinal mixed methods study of the experiences of students and their learning outcomes, tracing the experiences and educational outcomes of first year students studying in RCCs, to their subsequent enrollment in degree programs in Russia, to their return to their home countries. Another project could be a qualitative study from the perspective of teachers, instructors, and staff in RCCs, which could be very valuable in understanding the institutional perspectives, beyond my initial lens of State and Student. Also, the theoretical framework designed for the study could be applied to other cultural centers, such as Chinese Confucius Institutes. This topic has the potential to be examined beyond education research, in the fields of political science, international relations, peace studies, Black studies, and sociology.

Conclusions and Implications

"In the 1960s a joke went around Budapest about a man buying tea. When asked: which tea do you want: Russian or Chinese? he replied: I will have coffee instead." (Maru, 2017)

One of the main goals of American higher education is to prepare adults to participate in democratic citizenship, equipping students with the knowledge, resources, and tools to disseminate information, critique unjust power structures, and encourage the sound sustainability of society. However these values transcend borders with global implications, as Keller found that education has a marked impact on political interest and support for government in Kenyan students (Keller, 1980). Similarly, Zhang and Fagan (2016) highlight how educated young people in China are more likely to express civic expression and political intention than their non-educated peers (Zhang & Fagan, 2016).

There is extensive literature on Chinese Confucius Institutes with speculation that these schools apply ideological pressure to their foreign students. As discussed in this chapter, there is also extensive literature on Soviet education programs in Africa, but after the 1991 collapse, the Global North was less interested in the intrigue of the Cold War. Viewing higher education through the lens of critical geopolitics yields an understanding that the dominant idea that Global South seeks to immigrate to the U.S. (or the U.K.) is waning.

Gone are the days of superpowers, East-West and North-South dichotomies. The future of international education is that of a multipolar world, with both Student and State engaging in strategic maneuvering to benefit their own agendas. This work highlights the historical context of the present higher education development agenda wielded by the Russian Federation in Africa. In examining this area of study, the significance of Russia's geopolitical maneuvering points to a future of Chinese engagement.

When Chinese Vice Premier Xiao Deng addressed the United Nations General Assembly in 1974, he revealed Mao Zedong's now famous Three Worlds Theory, categorizing the U.S. and the USSR as the First World, Japan, Canada, Europe and Australia as the Second World, and Africa, continental Asia, and Latin America as the Third World (Yee, 1983). This theory is an application of critical geopolitics, moving beyond Alfred Sauvy's (1952) Three-World Model which categorizes the U.S. and its allies as the First, the USSR and China as the Second, and Africa, Latin America, and all of the countries in the Non-Aligned Movement of the Cold War as the Third World (Pletsch, 1981). This reframing: that of China sharing the third world with Africa, was the catalyst for the current, voracious economic and political agenda between the two regions (Gillespie, 2004). The establishment of the Forum on China-Africa Cooperation (FOCAC) in 1996 brokered official trade partnerships and relations between China and 53 African countries and the Commission of the African Union, solidified the following decades of influence (Kelly, 2017).

Much has been written on the neocolonial practices of predatory lending and exploitation of natural resources, and monopoly over industries wielded by China in Africa (Alden, 2005; Antwi-Boateng, 2017; Kelly, 2017; Lumumba-Kasongo, 2011; Mathews, 2021; Sartre, 2005). However this is not necessarily a deterrent for African students weighing the options for their futures. One might consider the old adage "if you can't beat 'em, join 'em" in this situation. Considering China's recent decision to increase the number of scholarships to Africa students (Ferdjani, 2012; Haugen, 2013; Monitor, 2021) as well as the Zambian RCC students interviewed in this study and their interest in China and Russia over the West—this book has implications for both policy and practice by the United States government.

In the past, the United States had the largest diplomatic presence in the world, but as of 2019, has been superseded by China. China has 276 embassies, consulates and diplomatic offices around the world, closely followed by the United States, France, Japan and Russia (*China now has more diplomatic posts than any other country*, 2019). However, there are many African people that are leery of the neocolonial behaviors China exhibits. For example:

> Chimuka Singuwa, a 23-year-old Zambian who is working towards a master's degree in international relations and diplomacy at RUDN, said he had the opportunity to study in Russia or China. He chose Moscow on the advice of his grandfather, who was a student at RUDN in the 1970s. Singuwa said he had no regrets about passing up on a Chinese education. "I'm kind of against the 'takeover' of Africa by the Chinese," he said, pointing to the debt Zambia has accumulated with China, its main investor. (*As Kremlin Scrambles for Africa*, 2019)

The Russian Federation and the prior Soviet Union engaged in educational development in Africa throughout the 20th century, including scholarship programs for African students to attend Russian universities, short-term cultural exchange programs, and technical training in-country, to teach mechanics, pilots, and engineers to replace the colonists who left after liberation. These programs were cultivated due to multiple motivations—some political, some ideological, and some ethical. While the literature provides a survey of all of the Soviet programming, this is marked gap on the topic of new institutions, post 1991. As goals changed from one regime to the next, so did educational practices.

Indeed, informal higher education institutions like Russian Cultural Centers are an example of a fifty-year trend of prioritizing the educational development of the non-university sector in Africa, rather than bolstering formal higher education (Omwami, 2013). Although it is a difficult battle against the vast neoliberal pressures weighing on higher education, there is room to assess U.S. policy and practice as a result of this research inquiry. Keller (2012) describes U.S. foreign policy in Africa as one of "selective engagement" with the U.S. only intervening or engaging when their geopolitical interests are at stake (Keller, 2012). If the United States is to thrive sustainably in the coming decades, simultaneously with Russia and China, we must aim beyond narrow self-serving interests, beyond battling superpowers to consider the multipolar world of this millennium.

References

Alden, C. (2005). China in Africa. *Survival*, 47(3), 147–164.

Antwi-Boateng, O. (2017). New World Order Neo-Colonialism: A Contextual Comparison of Contemporary China and European Colonization in Africa. *Journal of Pan African Studies*, 10(2), 177–195.

As Kremlin scrambles for Africa, Moscow University Eyes Soft Power. (2019, October 20, 2019). https://www.france24.com/en/20191020-as-kremlin- scrambles-for-africa-moscow-university-eyes-soft-power

Bernard S. Morris (1994) The end of ideology, the end of Utopia, and the end of history --On the occasion of the end of the U.S.S.R., History of European Ideas, 19:4–6, 699–708, DOI: 10.1016/0191-6599(94)90053-1

China Now Has More Diplomatic Posts Than Any Other Country. (2019). BBC News. Retrieved July 16, 2021 from https://www.bbc.com/news/world-asia-china-50569237

FBI. (1959). *William E.B. Dubois*. Federal Bureau of Investigation. Retrieved from https://vault.fbi.gov/E.%20B.%20%28William%29%20Dubois/E.%20B.%20%28William%29%20Dubois%20Part%205%20of%205

Ferdjani, H. 2012. African students in China: An exploration of increasing numbers and their motivations in Beijing. Research reports. Centre for Chinese Studies: Stellenbosch University. http://hdl.handle.net/10019.1/70764

Getachew, S. (2018, March 10). https://www.thereporterethiopia.com/article/ethiopia-russia-enter-new-frontier. *The Reporter*. https://www.thereporterethiopia.com/article/ethiopia-russia-enter-new-frontier

Gillespie S. (2014). *South-south transfer : a study of sino-african exchanges*. Taylor and Francis. Retrieved October 22 2023 from https://public.ebookcentral.proquest.com/choice/publicfullrecord.aspx?p=1595062.

Giorgis, E. W. (2020, June 4). Viewpoint: Race and Race Matters: A Personal Reflection. *Addis Standard*. https://addisstandard.com/viewpoint-race-and-race-matters-a-personal-reflection/

Hedenskog, J. (2018). *Russia Is Stepping up Its Military Cooperation in Africa* Security Policy, Issue. file:///Users/hopemccoy/Downloads/FOIMEMO6604.pdf

Keller, E. J. (2012). Meeting the Challenges of Strategic and Human Security Interests in US–Africa Relations, or the Orphaning of 'Soft Power'? *Africa Review*, 4(1), 1–16.

Kelly, R. (2017). How China's Soft Power Is Building a Neo-Colonial System in Africa. *Ketalag Media*, October, 9.

Knight, K. (2006). Transformations of the Concept of Ideology in the Twentieth Century. *American Political Science Review*, 100(4), 619–626.

Kochetkova, I. (2009). *The Myth of the Russian Intelligentsia: Old Intellectuals in the New Russia* (Vol. 62). Routledge.

Kulenova, A. (2019). *Africa: The New Frontier of Russian Influence*. The McGill International Review. Retrieved July 4, 2021 from https://www.mironline.ca/africa-the-new-frontier-of-russian-influence/

Light, M. (2003). International relations of Russia and the commonwealth of independent states. *Eastern Europe, Russia and Central Asia*.

Lipset, S. M. (2017) The end of ideology?. In: Ideology, pp. 47–66. Routledge.

Maru, M. T. (2017, July 1). The Rift in the GCC and Diplomatic Responses from the IGAD Region. *The Reporter*. https://www.thereporterethiopia.com/content/rift-gcc-and-diplomatic-responses-igad-region

Nawakwi Questions Govt's Decision to Buy Useless, Expensive Helicopters from Russia. (2015). https://www.zambiawatchdog.com/nawakwi-questions-govts-decision-to-buy-useless-expensive-helicopters-from-russia/

Omwami, E. (2013). Non-University Sector Reform. *Journal of Comparative & International Higher Education*, 5(Spring), 5–8.

Russell, M., & Pichon, E. (2019). Russia in Africa. A New Arena for Geopolitical Competition. *Briefing, European Parliamentary Research Service*, 81–12.

Pletsch, C. E. (1981). The Three Worlds, or the Division of Social Scientific Labor, Circa 1950-1975. *Comparative Studies in Society and History*, 23(4), 565–590. http://www.jstor.org/stable/178394

Tkachenko, V. (2018, April 7). Ethiopia and Russia: Kindred Spirits [Interview]. The Reporter. https://www.thereporterethiopia.com/article/ethiopia-and-russia-kindred-spirits

Yee, H. S. (1983). The Three World Theory and Post-Mao China's Global Strategy. *International Affairs (Royal Institute of International Affairs 1944–)*, 59(2), 239–249.

Young R. J. C. (2001). *Postcolonialism an historical introduction*. John Wiley & Sons Incorporated. Retrieved October 22 2023 from https://public.ebookcentral.proquest.com/choice/PublicFullRecord.aspx?p=7104509.

Young, K. R., & Green, D. S. (1972). Harbinger to Nixon: WEB Du Bois in China. *Negro History Bulletin*, 35(6), 125–128.

Zhang, C., & Fagan, C. (2016). Examining the Role of Ideological and Political Education on University Students' Civic Perceptions and Civic Participation in Mainland China: Some Hints from Contemporary Citizenship Theory. *Citizenship, Social and Economics Education*, 15(2), 117–142. https://doi.org/10.1177/2047173416681170

Rochelle Brock & Cynthia Dillard
Executive Editors

Black Studies and Critical Thinking is an interdisciplinary series which examines the intellectual traditions of and cultural contributions made by people of African descent throughout the world. Whether it is in literature, art, music, science, or academics, these contributions are vast and far-reaching. As we work to stretch the boundaries of knowledge and understanding of issues critical to the Black experience, this series offers a unique opportunity to study the social, economic, and political forces that have shaped the historic experience of Black America, and that continue to determine our future. Black Studies and Critical Thinking is positioned at the forefront of research on the Black experience, and is the source for dynamic, innovative, and creative exploration of the most vital issues facing African Americans. The series invites contributions from all disciplines but is specially suited for cultural studies, anthropology, history, sociology, literature, art, and music.

Subjects of interest include (but are not limited to):

- Education
- Sociology
- History
- Media/Communication
- Religion/Theology
- Women's Studies
- Policy Studies
- Advertising
- African American Studies
- Political Science
- LGBT Studies

For additional information about this series or for the submission of manuscripts, please contact Dr. Brock (University of North Carolina at Greensboro) at r_brock@uncg.edu or Dr. Dillard (University of Georgia) at cdillard@uga.com.

To order other books in this series, please contact our Customer Service Department:

peterlang@presswarehouse.com (within the U.S.)
order@peterlang.com (outside the U.S.)

Or browse online by series at www.peterlang.com.

www.ingramcontent.com/pod-product-compliance
Lightning Source LLC
Chambersburg PA
CBHW061720300426
44115CB00014B/2768